EXPERIMENTAL MECHANICS OF FIBER REINFORCED COMPOSITE MATERIALS

Revised Edition

by

James M. Whitney
Air Force Materials Laboratory
Wright-Patterson AFB

Isaac M. Daniel
Illinois Institute of Technology
Research Institute, Chicago, IL

R. Byron Pipes
University of Delaware
Newark, Delaware

Published by
THE SOCIETY FOR EXPERIMENTAL MECHANICS
Brookfield Center, Connecticut

Prentice-Hall, Inc., Englewood Cliffs, New Jersey 07632

The Monograph Committee of SEM (formerly SESA) developed this book, which is believed to be the first of its type, to service both the composites and experimental mechanics communities. As a result, the publication is expected to be well received by both groups.

The Monograph Committee wishes to thank the authors who voluntarily gave their time to produce this volume. Further, we would like to recognize the efforts of the numerous technical reviewers who donated their expertise to this project. Credit is also due K.A. Galione and M.E. Yergin of SEM for production coordination assistance, and, finally, our sincere thanks to Prof. S.E. Swartz, who initiated the "Experimental Mechanics of Fiber Reinforced Composite Materials" project as chairman of the Monographs Committee from 1978-1980.

C.W. Bert, Chairman, Monographs Committee 1980-82
J.W. Phillips, Vice Chairman

ISBN 0-912053-01-1

SEM Monograph No. 4, Revised Edition

14 Fairfield Drive
Brookfield Center, CT 06805
Printed in U.S.A.

First edition, 1982
Revised edition, 1984

CONTENTS

EXPERIMENTAL MECHANICS OF FIBER REINFORCED COMPOSITE MATERIALS

Revised Edition

SOCIETY FOR EXPERIMENTAL MECHANICS MONOGRAPH

1 INTRODUCTION

With the increasing use of composite materials in structural applications has come a corresponding increase in the need for experimental data. Thus, composite material test methodology has received considerable attention in recent years. Many of the test methods used to characterize metals have been applied to fiber reinforced materials. The heterogeneous, anisotropic nature of fiber reinforced composites requires, however, that any test method borrowed from metallic technology be carefully scrutinized before being directly applied to the characterization of composite materials.

The need to critically evaluate composite test methods has increased in recent years due to the complex nature of data required for design consideration. In the early stages of composite technology development, test methods revolved around measuring strength and stiffness under simple tension, compression, and shear loads. With increasing use of composite materials in structural design came the need for more advanced characterization such as notched strength, strength under multi-axial loading, and determination of fatigue life. A number of surveys of a variety of composite materials test methods can be found in the literature.[1-5]

The objective of this monograph is to review state-of-the-art test methods in fiber reinforced composites from an applied mechanics point of view. Such an approach is appropriate in light of the complex nature of composite materials. Thus, the monograph attempts to present composite materials test methodology on a rational basis, rather than simply presenting a survey of test methods. As a result, all known test methods used in conjunction with a specific property measurement are not necessarily discussed. The methods presented in detail are considered, by the authors, to be the currently most viable approaches based on theoretical considerations, personal experience, and knowledge of other workers' experiences. Some test methods, such as the short-beam shear method, which are not the most ideal from a theoretical standpoint are discussed in detail because of the overwhelming popularity of the method. Many of the methods presented are current ASTM standards or are currently being considered for ASTM standardization. Test methods for measuring certain properties, such as impact resistance, are not well developed at the present time and as such are not considered to be within the scope of this monograph.

In choosing test methods which provide the optimum or "best" approach to measuring a particular mechanical property, a definition is needed of what constitutes an ideal test method. An ideal test method is one in which a known state of uniform stress and strain can be introduced throughout the test specimen. In the real world, however, the experimentalist is forced to accept as "optimum" methods those in which a known state of uniform stress and strain is obtained in the specimen gage section while stress concentrations are minimized in the load introduction region. The problem of defining a region of uniform stress and strain in composite materials is complicated by the breakdown of St. Venant's principle in the testing of highly orthotropic materials.[6] In particular, it may take a much longer specimen length to dissipate the effect of load introduction in an orthotropic material than in the case of an isotropic material. This must be a consideration when designing test specimen geometry.

Although much of the work presented in this monograph is applicable to a wide variety of composites there is a definite slant toward materials for aerospace applications. This is primarily due to the fact that most of the technology in recent years has been heading in that direction.

A review of the current state-of-the-art in mechanics of composite materials is presented in Chapter 2. The theory presented is based on Hooke's law. Although some nonlinear behavior is observed in engineering composites, especially in the off-axis and shear response of unidirectional material, most of the current state-of-the-art in composite mechanics is based on linear elastic behavior. Linear stress–strain behavior is usually observed at room temperature parallel and transverse to the fiber direction in a unidirectional composite, with considerable nonlinearity observed in the shear response. Much of the nonlinear stress–strain response observed in laminates at room temperature is associated with damage (usually matrix cracking). This type of behavior is discussed in Chapter 4 in conjunction with first ply failure. At elevated temperatures nonlinear and time dependent stress–strain behavior can be observed in conjunction with matrix dominated properties. Formal consideration of nonlinear and/or time dependent stress–strain behavior is considered, however, to be beyond the scope of an initial monograph on experimental mechanics of composite materials. Such behavior could form the basis of another monograph in this area. Concepts presented in Chapter 2 form the basis of test method analysis in later chapters.

The experimentalist's tools such as strain gages, photoelastic coatings, moiré fringes, and holographic techniques, are discussed in Chapter 3. Although these techniques are familiar in metals technology, special consideration must be given to their applicability to fiber-reinforced composites.

Specific test methods are discussed in Chapter 4 with special emphasis on mechanical properties, rather than physical properties. With the exception of inplane shear, this chapter provides a relatively complete survey of methods used in conjunction with the properties considered. Shear behavior is very difficult to determine experimentally, as it is virtually impossible to design a test specimen which will produce pure shear in the gage section without introducing significant stress concentrations in the remainder of the test specimen. As a result, numerous shear tests have been developed or proposed. The methods chosen for discussion are based on theoretical considerations, current usage, and current acceptance. Other methods not covered that are of historical significance include the Douglas split-ring test,[7] the cruciform sandwich beam test,[8] the picture-frame shear test,[9] and the sheet torsion test.[10]

Effects of moisture on composites and corresponding experimental ramifications are discussed in Chapter 5. Current concern over the effect of moisture on composites having polymeric matrices (especially epoxy resins) provides an important need for such material to be included in this monograph. Some data reported in the literature on the effect of moisture on mechanical properties have been obtained incorrectly. A basic understanding of the mechanisms associated with moisture diffusion is necessary in order to prevent erroneous data and corresponding erroneous conclusions from being reported in the technical literature.

This monograph should be of use to students, researchers, and engineers who have any interest in the experimental aspects of fiber reinforced composites. Furthermore, it should provide an incentive to evaluate test methods in terms of sound principles of applied mechanics rather than blindly applying a method developed for homogeneous, isotropic materials without questioning its applicability to heterogeneous, anisotropic composites.

REFERENCES

1. Lenoe, E.M., "Testing and Design of Advanced Composite Materials," Journal of the Engineering Mechanics Division, *American Society of Civil Engineers,* Vol. 96, No. EM6, 809-823 (Dec. 1970).

2. Prosen, S.P., "Composite Materials Testing," *Composite Materials: Testing and Design,* ASTM STP 460, American Society for Testing and Materials, 5-12 (1969).
3. Bert, C.W., "Experimental Characterization of Composites," Chapter 9, *Structural Design and Analysis,* Part II, C.C. Chamis, editor, *Composite Materials,* Vol. 8, L.J. Broutman and R.H. Krock, editors, Academic Press, New York, 73-133 (1975).
4. Chiao, T.T. and Hamstad, M.A., "Testing of Fiber Composite Materials," *Proceedings of the 1975 International Conference on Composite Materials,* Vol. 2, The Metallurgical Society of AIME, 884-915 (1976).
5. Agarwal, Bhagwan D. and Broutman, Lawrence J., "Experimental Characterization of Composites," Chapter 8, *Analysis and Performance of Fiber Composites,* Wiley-Interscience Publications, John Wiley and Sons, New York (1980).
6. Choi, I. and Horgan, C.O., "Saint-Venant's Principle and End Effects in Anisotropic Elasticity," *Journal of Applied Mechanics,* Vol. 44, No. 3, 424-430 (Sept. 1977).
7. Greszczuk, L.B., "Douglas Ring Test for Shear Modulus Determination of Isotropic and Composite Materials," *Proceedings, 23rd Technical Conference, Reinforced Plastics/Composites Division, Society of the Plastics Industry,* Section 17-D (Feb. 1968).
8. Waddoups, M.E., "Characterization and Design of Composite Materials," *Composite Materials Workshop,* S.W. Tsai, J.C. Halpin and N.J. Pagano, editors, Technomic Publishing Co., Stamford, CT, 254-308 (1968).
9. Penton, A.P., "A New Device for Determining Shear Properties of Reinforced Plastics," *SPE Journal,* Vol. 16, 1246-1247 (Nov. 1960).
10. Lempriere, B.M., Fenn, R.W., Jr., Crooks, D.D. and Kinder, W.C., "Torsion Testing for Shear Modulus of Thin Orthotropic Sheet," *AIAA Journal,* Vol. 7, 2341-2342 (Dec. 1969).

2

MECHANICS OF COMPOSITE MATERIALS

2.1 NOMENCLATURE

A composite material is defined as a material containing two or more distinct phases on a macroscopic scale. We will be dealing with materials which contain a fiber reinforcing material supported by a binder or matrix material. In most engineering applications, these materials are used in the form of laminates which consist of layers of various composite materials. These individual layers are constructed of unidirectional continuous fibers, unidirectional discontinuous fibers, or woven cloth. A clearly defined nomenclature is necessary to identify material properties relative to a principal material coordinate system within each layer of a laminate and to identify laminate stacking geometry. In addition, since we will be continually dealing with stress, strain, and elastic properties of composite materials, it is important to establish a consistent nomenclature for these quantities.

2.1.1 Components of Stress and Strain

Consider an orthogonal cartesian coordinate system x_1, x_2, and x_3. The tensorial stress is denoted by σ_{ij} $(i,j = 1,2,3)$ where the first subscript indicates a direction normal to the area on which the stress component acts and the second subscript indicates the direction of the stress component. Thus, normal stress is indicated by repeated subscripts, while a shear stress is indicated by mixed subscripts. Tensorial strains are denoted by ϵ_{ij} $(i,j = 1,2,3)$ with the subscripts defined in a manner analogous to the stress tensor subscripts.

Both the stress and strain tensors are symmetric, that is

$$\sigma_{ij} = \sigma_{ji}, \qquad \epsilon_{ij} = \epsilon_{ji}, \qquad (i,j = 1,2,3) \tag{2.1}$$

Thus, the number of stress and strain components are reduced from nine to six. In practical applications, engineering notation is preferred for shear components of stress and strain rather than the tensorial notation. In particular, engineering notation is given by

$$\sigma_{ij} = \tau_{ij}, \qquad \epsilon_{ij} = \frac{\gamma_{ij}}{2}, \qquad (i \neq j) \tag{2.2}$$

The following contracted notation is also used

$$\sigma_{ii} = \sigma_i, \qquad \epsilon_{ii} = \epsilon_i \qquad (i = 1,2,3) \tag{2.3}$$

$$\tau_{23} = \sigma_{23} = \sigma_4, \qquad \gamma_{23} = 2\epsilon_{23} = \epsilon_4 \tag{2.4}$$

$$\tau_{13} = \sigma_{13} = \sigma_5, \qquad \gamma_{13} = 2\epsilon_{13} = \epsilon_5 \tag{2.5}$$

$$\tau_{12} = \sigma_{12} = \sigma_6, \qquad \gamma_{12} = 2\epsilon_{12} = \epsilon_6 \tag{2.6}$$

2.1.2 Principal Material Coordinate System

Laminated composites are constructed of plies containing unidirectional fibers or woven cloth. An x_1, x_2 axis system parallel and transverse to the fibers, respectively, in a unidirectional layer is referred to as the principal material coordinate system. The orientation of the fibers in each ply relative to the x axis of the laminate is denoted by θ (see Fig. 2-1). In the case of woven cloth, the x_1 and x_2 directions correspond to the warp and fill directions of the fabric, respectively.

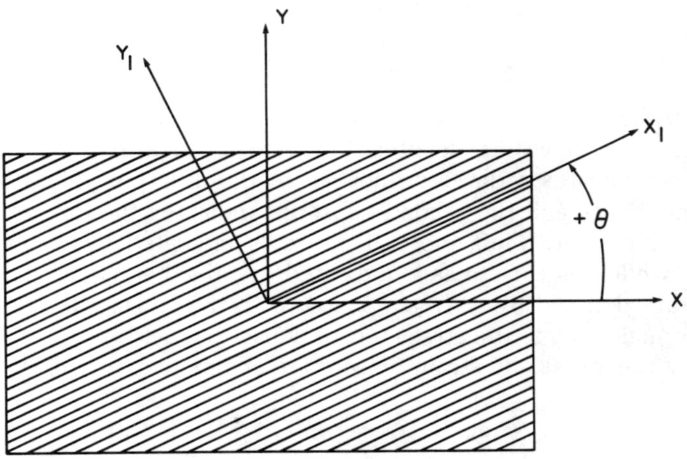

Fig. 2-1—Orientation of principal material axes in a unidirectional layer

2.1.3 Elastic Symmetry

The number of elastic constants which describe a unidirectional composite is a function of the packing geometry. Typical packing geometries are shown in Fig. 2-2. For the rectangular packing in Fig. 2-2a, nine elastic constants are necessary to describe the behavior. Three Young's moduli E_1, E_2, and E_3 parallel to the axes x_1, x_2, and x_3, respectively; three shear moduli G_{12}, G_{13}, and G_{23} associated with shear strains in the x_1x_2, x_1x_3, and x_2x_3 planes respectively; and three major Poisson ratios ν_{12}, ν_{13}, and ν_{23}. The first subscript in the Poisson ratios denotes the direction of uniaxial tensile stress, while the second subscript denotes the direction in which the contraction is measured. It will be shown in later sections of this chapter that the minor Poisson ratios ν_{21}, ν_{31}, and ν_{32} are not independent constants.

For the case of square packing, Fig. 2-2b, it is easily shown from symmetry considerations that

$$E_3 = E_2, \qquad G_{13} = G_{12}, \qquad \nu_{13} = \nu_{12} \tag{2.7}$$

Thus, the number of independent elastic constants is reduced to six. In the case of hexagonal packing, Fig. 2-2c, the x_2x_3 plane is a plane of isotropy in which the following classic isotropic relationship is valid

$$G_{23} = \frac{E_3}{2(1 + \nu_{23})} \tag{2.8}$$

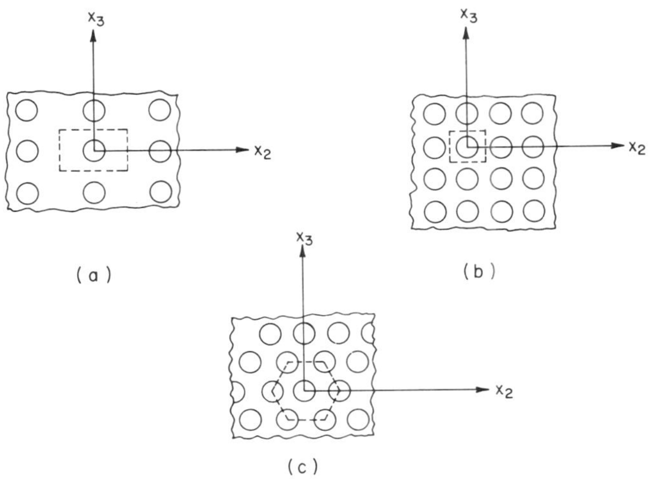

Fig. 2-2—Unidirectional fiber packing geometry: (a) rectangular array, (b) square array, (c) hexagonal array

Such a material is referred to as transversely isotropic with the number of independent elastic constants being reduced to five. Glass fibers and graphite fibers tend to nest during fabrication, resulting in hexagonal packing and five independent elastic constants.

For the remainder of this book the inplane behavior of the material will be of primary concern. Thus, regardless of packing geometry, only the inplane ply elastic constants E_1, E_2, G_{12} and ν_{12} are of practical interest.

2.1.4 Laminate

A laminated composite is constructed of a number of plies of unidirectional or woven fabric composites stacked at various angles relative to the x axis of the laminate. In order to describe the stacking geometry of a laminated composite, it is necessary to have a code which describes a laminate uniquely. For the case of equal ply thickness, the stacking sequence can be described by simply listing the ply orientations, θ, from top to bottom. Thus, the notation $[0/90/0]_T$ uniquely describes a three layer laminate. The subscript T denotes that the sequence accounts for the total number of layers. If a laminate is symmetric, as in the case $[0/90_2/0]_T$, the notation can be abbreviated by using $[0/90]_s$. The subscript s denotes that the stacking sequence is repeated symmetrically about the laminate centerline. The subscript *2* in total stacking sequence notation is used to show that the 90 degree ply is repeated. Angle-ply laminates are denoted by $[0/45/-45]_s$, which can be shortened to $[0/\pm 45]_s$. For laminates with repeating sets of plies, such as $[0/\pm 45/0/\pm 45]_s$, a shorter notation takes the form $[0/\pm 45]_{2s}$. If a layer is split at the centerline in a symmetric laminate, a bar is used over the center ply to denote the split. For example, the laminate $[0/90/0]_T$ can be abbreviated to $[0/\overline{90}]_s$.

The notation as discussed here is consistent with the Air Force Design Guide[1] and ASTM codes.

2.2 PREDICTION OF LAMINA PROPERTIES—MICROMECHANICS

The term "micromechanics" is used to describe analysis of unidirectional composites in which effective lamina properties are derived in terms of constituent material properties, geometric arrangement of the constituents, and relative volume content of the constituent materials. Such analysis is often complex, requiring approximate numerical methods in order to obtain solutions. A number of approximate closed form expressions have been developed, however, which are useful in estimating lamina properties from constituent material properties.

These relations will be presented for the case of orthotropic constituents. Since graphite fibers are orthotropic, such a generalization has practical value.

2.2.1 Young's Moduli

Young's modulus parallel to the fibers, E_1, can be estimated by a simple rule-of-mixtures for continuous filament composites,[2] i.e.,

$$E_1 = E_{1f} V_f + E_{1m}(1 - V_f) \tag{2.9}$$

where E_{1f}, E_{1m}, and V_f are the longitudinal Young's modulus of the fiber, longitudinal Young's modulus of the matrix, and the volume fraction of the fiber, respectively. For unidirectional composites in which the fibers are not continuous, but having constant length, b, eq (2.9) takes the form[3]

$$E_1 = \frac{E_{1m}[1 + 2(b/d)\eta_1 V_f]}{(1 - \eta_1 V_f)} \tag{2.10}$$

where d is the fiber diameter and

$$\eta_1 = \frac{(E_{1f}/E_{1m} - 1)}{[E_{1f}/E_{1m} + 2(b/d)]} \tag{2.11}$$

Note that as b/d becomes very large, eq (2.10) reduces the rule-of-mixtures as given by eq (2.9).

The transverse unidirectional Young's modulus, E_2, can be estimated for either continuous filament composites or discontinuous filament composites from the relationship[3-5]

$$E_2 \doteq \frac{E_{2m}(1 + \xi\eta_2 V_f)}{(1 - \eta_2 V_f)} \tag{2.12}$$

where

$$\eta_2 = \frac{(E_{2f}/E_{2m} - 1)}{(E_{2f}/E_{2m} + \xi)} \tag{2.13}$$

and E_{2f}, E_{2m}, and ξ are the transverse Young's modulus of the fiber, transverse Young's modulus of the matrix, and a fiber packing geometry factor, respectively. In particular,

$$\begin{aligned} \xi &= 2, \text{ square packing} \\ &= 1, \text{ hexagonal packing} \end{aligned}$$

2.2.2 Poisson's Ratios

The major Poisson's ratio, ν_{12}, of a unidirectional composite can be estimated by a simple rule-of-mixtures for continuous or discontinuous fibers.[2,3] Thus,

$$\nu_{12} = \nu_{12f} V_f + \nu_{12m}(1 - V_f) \tag{2.14}$$

where ν_{12f} and ν_{12m} are the major Poisson's ratio of the fiber and matrix, respectively.

As will be shown later, the minor Poisson's ratio, ν_{21}, is not an independent elastic property and can be determined from the relationship

$$\nu_{21} = \nu_{12} \frac{E_2}{E_1} \tag{2.15}$$

For transversely isotropic unidirectional composites, the transverse Poisson's ratio, ν_{23}, is determined from the relationship

$$\nu_{23} = \frac{2E_1K_2 - E_1E_2 - 4\nu_{12}^2K_2E_2}{2E_1K_2} \tag{2.16}$$

where K_2 is the plane strain bulk modulus. For continuous fiber reinforced unidirectional material, K_2 can be estimated as follows.[2]

$$K_2 = \frac{(K_{2f} + G_{23m})K_{2m} + (K_{2f} - K_{2m})G_{23m}\, V_f}{(K_{2f} + G_{23m}) - (K_{2f} - K_{2m})\, V_f} \tag{2.17}$$

where K_{2f} and K_{2m} are the plane strain bulk moduli for the fiber and matrix, respectively.

2.2.3 Shear Moduli

For either continuous or discontinuous filament unidirectional composites, the inplane shear modulus, G_{12}, can be estimated from the relationship[2,3]

$$G_{12} = \frac{G_{12m}[G_{12f}(1 + V_f) + G_{12m}(1 - V_f)]}{G_{12f}(1 - V_f) + G_{12m}(1 + V_f)} \tag{2.18}$$

where G_{12f} and G_{12m} are the inplane shear moduli of the fiber and matrix, respectively. For transversely isotropic materials, eqs (2.9, 12, 13, 16, 17) can be used in conjunction with eq (2.8) to estimate the transverse shear modulus G_{23}.

2.2.4 Expansional Strains

In polymeric matrix composites volume changes due to absorbed moisture from a humid environment are equally as important as volume change due to temperature. As a result, the concept of generalized expansional strains, ϵ_i^E, is often used in the analysis of these materials. In particular,

$$\epsilon_i^E = \alpha_i \Delta T + \beta_i \Delta C \qquad (i = 1,2,6) \tag{2.19}$$

where α_i and β_i are thermal expansion and moisture expansion coefficients, respectively. Temperature is denoted by T and moisture concentration by C.

Approximate micromechanics expressions for thermal expansion coefficients have been derived by Schapery[6] for isotropic constituents. The longitudinal coefficient α_1, for continuous filament composites is given by the relationship

$$\alpha_1 = \frac{\overline{E\alpha}}{\overline{E}} \tag{2.20}$$

where the bar denotes volume average rule-of-mixture relationships, i.e.,

$$\overline{E\alpha} = E_f \alpha_f V_f + E_m \alpha_m (1 - V_f) \tag{2.21}$$

$$\overline{E} = E_f V_f + E_m (1 - V_f) \tag{2.22}$$

with the fiber and matrix coefficients of thermal expansion denoted by α_f and α_m, respectively. For discontinuous fiber unidirectional composites, eq (2.20) is replaced by a modification of Schappery's work suggested by Halpin[3]

$$\alpha_1 = \overline{\alpha} + \frac{(E_1 - E_u)(\overline{E^{\alpha}} - \overline{E\alpha})}{E_1(\overline{E} - E_u)} \tag{2.23}$$

where

$$\overline{\alpha} = \alpha_f V_f + \alpha_m (1 - V_f) \tag{2.24}$$

$$E_u = \frac{E_f E_m}{E_f(1 - V_f) + E_m V_f} \tag{2.25}$$

and E_1 refers to the longitudinal modulus for discontinuous fibers as given by eq (2.10). Hashin[2] derived thermal expansion coefficients for the case of continuous orthotropic fibers in an orthotropic matrix. The longitudinal coefficient can be approximated by eq (2.20) with

$$\overline{E\alpha} = E_{1f}\alpha_{1f} V_f + E_{1m}\alpha_{1m}(1 - V_f) \tag{2.26}$$

$$\overline{E} = E_{1f} V_f + E_{1m}(1 - V_f) \tag{2.27}$$

where α_{1f} and α_{1m} denote longitudinal thermal expansion coefficients of the fiber and matrix, respectively.

The transverse thermal expansion coefficient, α_2, for both continuous and discontinuous filament unidirectional composites with isotropic constituents can be estimated from the relationship[6]

$$\alpha_2 = \alpha_f V_f(1 + \alpha_f) + \alpha_m(1 - V_f)(1 + \nu_m) - [\nu_f V_f + \nu_m(1 - V_f)] \frac{\overline{E\alpha}}{\overline{E}} \tag{2.28}$$

For orthotropic constituents, α_2 can be estimated by[2]

$$\alpha_2 = \alpha_{2f}(1 + \nu_{12f}\alpha_{1f}/\alpha_{2f}) V_f + \alpha_{2m}(1 + \nu_{12m}\alpha_{1m}/\alpha_{2m})(1 - V_f) - [\nu_{12f} V_f + \nu_{12m}(1 - V_f)] \frac{\overline{E\alpha}}{\overline{E}} \tag{2.29}$$

where α_{2f} and α_{2m} are the transverse expansion coefficients of the fiber and matrix, respectively, and $\overline{E\alpha}$ and $\overline{E}$ are given by the modified relationships displayed in eqs (2.26-27).

In the case of moisture expansion, the analytical procedures for deriving β_1 are exactly analogous to α_1 and α_2. Thus, eqs (2.20-29) are applicable with β's. In many practical applications, however, it is the matrix material which absorbs the moisture with the fiber being inert relative to moisture absorption. Such is the case with boron and graphite fibers. In these cases $\beta_f = \beta_{1f} = \beta_{2f} = 0.$ and considerable simplification is associated with the micromechanics expressions. In particular, for continuous fiber unidirectional composites eq (2.20) for orthotropic constituents becomes

$$\beta_1 = \beta_{1m} \frac{E_{1m}(1 - V_f)}{E_{1f} V_f + E_{1m}(1 - V_f)} \tag{2.30}$$

For discontinuous fibers, eq (2.23) becomes

$$\beta_1 = \beta_m(1 - V_f) \left[1 - \frac{(E_1 - E_u)(E_f - E_m) V_f}{E_1(E - E_u)}\right] \tag{2.31}$$

In the case of the transverse expansion coefficient, β_2, for both continuous and discontinuous fibers eq (2.28) reduces to

$$\beta_2 = \beta_m \frac{(1 - V_f)}{\overline{E}} \{(1 + \nu_m) V_f E_f + [1 - (1 + \nu_f) V_f] E_m\} \tag{2.32}$$

and for orthotropic constituents, eq (2.29) reduces to

$$\beta_2 = \beta_{2m} \frac{(1 - V_f)}{\overline{E}} \{(1 + \nu_{12m}\beta_{1m}/\beta_{2m})(E_{1f} V_f + E_{1m}(1 - V_f)$$

$$- [\nu_{12f} V_f + \nu_{12m}(1 - V_f)] E_{1m}\beta_{1m}/\beta_{2m}\} \tag{2.33}$$

where $\overline{E}$ is defined by eq (2.24).

2.3 STRESS AND STRAIN TRANSFORMATION—PLANE STRESS

Since the principal material coordinate system for the lamina does not generally coincide with the principal axes of a multi-directional laminate, it is necessary to transform both stress and strain components from one coordinate system to another.

2.3.1 Stress Transformation

If the components of stress at a point are known in one coordinate system, it is possible to obtain the stress components in any other coordinate system through tensor transformation of stress. If the stress components σ_x, σ_y, and τ_{xy} are known, σ_1, σ_2 and τ_{12} can be found by satisfying the following equations (see Fig. 2.3).

$$\begin{aligned} \sigma_1 &= m^2\sigma_x + n^2\sigma_y + 2mn\tau_{xy} \\ \sigma_2 &= n^2\sigma_x + m^2\sigma_y - 2mn\tau_{xy} \\ \tau_{12} &= -mn\sigma_x + mn\sigma_y + (m^2 - n^2)\tau_{xy} \end{aligned} \tag{2.34}$$

where $m = \cos\theta$ and $n = \sin\theta$.

2.3.2 Strain-Displacement Relations

Strains at a point in a continuum can be related to the displacement field of the continuum. If u and v are the x and y components of the displacement vector then the normal strain components in the x-y coordinate system are given as:

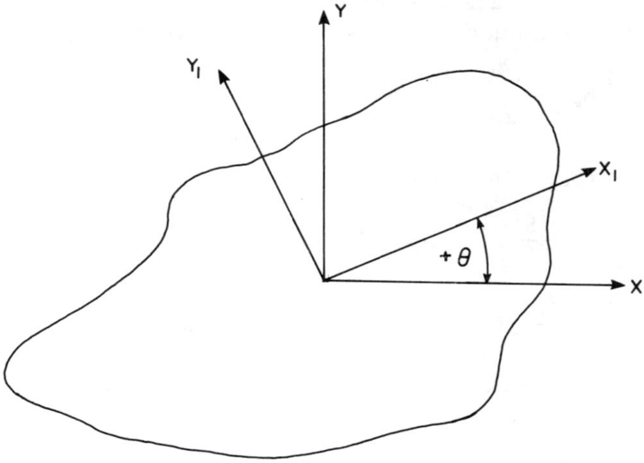

Fig. 2-3—Rotation of axes in *x-y* plane

$$\epsilon_x = \partial u/\partial x$$
$$\epsilon_y = \partial u/\partial y \tag{2.35}$$

Two versions of shear strain commonly referred to as "engineering" and tensorial shear strain are given as follows:

$$\gamma_{xy} = \frac{\partial u}{\partial y} + \frac{\partial v}{\partial x} \quad \text{(engineering)} \tag{2.36a}$$

$$\epsilon_{xy} = \frac{1}{2}\left(\frac{\partial u}{\partial y} + \frac{\partial v}{\partial x}\right) \quad \text{(tensorial)} \tag{2.36b}$$

2.3.3 Strain Transformation

The distinction between engineering and tensorial shear strain is made by their properties of transformation. The elasticity shear strain deformation allows strain to obey tensorial transformation and thus transform identically to stress. If the strain components ϵ_x, ϵ_y, and ϵ_{xy} are known, ϵ_1, ϵ_2, and ϵ_{12} can be found through transformation

$$\epsilon_1 = m^2\epsilon_x + n^2\epsilon_y + 2mn\epsilon_{xy}$$
$$\epsilon_2 = n^2\epsilon_x + m^2\epsilon_y - 2mn\epsilon_{xy} \tag{2.37}$$
$$\epsilon_{12} = -mn\epsilon_x + mn\epsilon_y + (m^2 - n^2)\epsilon_{xy}.$$

If, however, it is preferred to work with the engineering definition of shear strain then the non-tensorial transformation is given as follows:

$$\begin{aligned}
\epsilon_1 &= m^2\epsilon_x + n^2\epsilon_y + mn\gamma_{xy} \\
\epsilon_2 &= n^2\epsilon_x + m^2\epsilon_y - mn\gamma_{xy} \\
\gamma_{12} &= -2mn\epsilon_x + 2mn\epsilon_y + (m^2 - n^2)\gamma_{xy}
\end{aligned} \tag{2.38}$$

where $m = \cos\theta$ and $n = \sin\theta$.

2.4 STRESS–STRAIN RELATIONS

In order to perform engineering analysis the heterogeneous lamina consisting of fibrous and matrix phases is modeled as a homogeneous and orthotropic medium which possesses three planes of elastic symmetry. Further, the lamina is assumed to be subjected to a state of plane stress.

2.4.1 Engineering Material Properties of a Lamina

In order to develop the stress–strain relation for the lamina it is necessary to define the engineering material properties. The properties include Young's modulus in the fiber direction, E_1; transverse to the fiber direction, E_2; major Poisson's ratio, ν_{12}; minor Poisson's ratio, ν_{21}; and shear modulus, G_{12}.

$$E_1 = \frac{\sigma_1}{\epsilon_1}, \quad \text{when} \quad \sigma_2 = \tau_{12} = 0 \tag{2.39}$$

$$E_2 = \frac{\sigma_2}{\epsilon_2}, \quad \text{when} \quad \sigma_1 = \tau_{12} = 0 \tag{2.40}$$

$$\nu_{12} = -\frac{\epsilon_2}{\epsilon_1}, \quad \text{when} \quad \sigma_2 = \tau_{12} = 0 \tag{2.41}$$

$$\nu_{21} = -\frac{\epsilon_1}{\epsilon_2}, \quad \text{when} \quad \sigma_1 = \tau_{12} = 0 \tag{2.42}$$

$$G_{12} = \frac{\tau_{12}}{\gamma_{12}}, \quad \text{when} \quad \sigma_1 = \sigma_2 = 0 \tag{2.43}$$

Having defined the engineering material properties it is now possible to write the stress–strain relations for an orthotropic material in terms of the engineering material properties.

$$\epsilon_1 = \frac{\sigma_1}{E_1} - \nu_{21}\frac{\sigma_2}{E_2} + (0)\tau_{12}$$

$$\epsilon_2 = -\frac{\nu_{12}\sigma_1}{E_1} + \frac{\sigma_2}{E_2} + (0)\tau_{12} \tag{2.44}$$

$$\gamma_{12} = (0)\sigma_1 + (0)\sigma_2 + \frac{\tau_{12}}{G_{12}}$$

It is important to note that eq (2.44) shows that the shear and normal responses are totally uncoupled in principal material coordinate system x_1, x_2.

2.4.2 Compliance Properties

It is often desirable to express the stress–strain relation for the lamina in matrix form. Hence if the shear stress and strain (engineering) components are written in contracted notation, eq (2.6), then eq (2.44) can be rewritten as follows:

$$\epsilon_i = S_{ij}\sigma_j \qquad (i,j = 1,2,6) \tag{2.45}$$

where double subscript indicates a summation. Written in expanded form eq (2.45)

$$\epsilon_1 = S_{11}\sigma_1 + S_{12}\sigma_2 + S_{16}\sigma_6$$

$$\epsilon_2 = S_{21}\sigma_1 + S_{22}\sigma_2 + S_{26}\sigma_6 \tag{2.46}$$

$$\epsilon_6 = S_{61}\sigma_1 + S_{62}\sigma_2 + S_{66}\sigma_6$$

Comparing eqs (2.44) and (2.46) it is now possible to determine the S_{ij} or compliance constants in terms of the engineering material properties.

$$S_{11} = 1/E_1 \qquad S_{12} = -\nu_{21}/E_2$$

$$S_{21} = -\nu_{12}/E_1 \qquad S_{66} = 1/G_{12}$$

$$S_{22} = 1/E_2 \qquad S_{26} = 0 \tag{2.47}$$

$$S_{61} = 0 \qquad S_{62} = 0$$

$$S_{16} = 0$$

For that class of fibrous composites wherein tensile and compressive elastic properties are identical, the compliance matrix, S_{ij}, is symmetric. Hence, $S_{12} = S_{21}$:

$$\nu_{21}/E_2 = \nu_{12}/E_1 \tag{2.48}$$

Equation (2.48) is known as the reciprocity relation. Thus the minor Poisson's ratio can be written in terms of the other engineering material properties:

$$\nu_{21} = \frac{E_2}{E_1}\nu_{12} . \tag{2.49}$$

2.4.3 Stiffness Properties

The stress–strain relation given in eq (2.44) presents the strain components as functions of engineering material properties and stress components. This relation may be inverted to yield stress components as a function of material properties and strain components. In this form the stress–strain relation is known as the stiffness relation.

$$\begin{aligned}
\sigma_1 &= \frac{E_1\epsilon_1}{(1-\nu_{12}\nu_{21})} + \frac{\nu_{21}E_1\epsilon_2}{(1-\nu_{12}\nu_{21})} + (0)\epsilon_6 \\
\sigma_2 &= \frac{\nu_{12}E_2\epsilon_1}{(1-\nu_{12}\nu_{21})} + \frac{E_2\epsilon_2}{(1-\nu_{12}\nu_{21})} + (0)\epsilon_6 \\
\sigma_6 &= (0)\epsilon_1 + (0)\epsilon_2 + G_{12}\epsilon_6
\end{aligned} \tag{2.50}$$

Again it should be noted that the constants which couple shear and normal response are zero in the principal coordinate system. In matrix form eq (2.50) becomes

$$\sigma_i = Q_{ij}\epsilon_j \quad (i,j = 1,2,6) \tag{2.51}$$

where the Q_{ij} are known as the stiffness constants. Equation (2.51) can be written in expanded form as follows:

$$\begin{aligned}
\sigma_1 &= Q_{11}\epsilon_1 + Q_{12}\epsilon_2 + Q_{16}\epsilon_6 \\
\sigma_2 &= Q_{21}\epsilon_1 + Q_{22}\epsilon_2 + Q_{26}\epsilon_6 \\
\sigma_6 &= Q_{66}\epsilon_1 + Q_{62}\epsilon_2 + Q_{66}\epsilon_6 .
\end{aligned} \tag{2.52}$$

Comparing eqs (2.52) and (2.50) allows identification of the Q_{ij} in terms of the engineering material properties.

$$
\begin{aligned}
Q_{11} &= \frac{E_1}{1-\nu_{12}\nu_{21}} & Q_{12} &= \frac{\nu_{21}E_1}{1-\nu_{12}\nu_{21}} \\
Q_{21} &= \frac{\nu_{12}E_2}{1-\nu_{12}\nu_{21}} & Q_{22} &= \frac{E_2}{1-\nu_{12}\nu_{21}} \\
Q_{66} &= G_{12} & Q_{26} &= 0 \\
Q_{16} &= 0 & Q_{62} &= 0 \\
Q_{61} &= 0
\end{aligned}
\qquad (2.53)
$$

The reciprocity relation, eq (2.48), in conjunction with eq (2.53) yields $Q_{21} = Q_{12}$. This symmetry is assured since the stiffness matrix is the inverse of a symmetric compliance matrix.

2.4.4 Arbitrary Coordinate System

Since the components of stress and strain may be transformed from one coordinate system to another it is possible to establish the stress–strain (compliance or stiffness) relation in any coordinate system. For example, the compliance relation in the x-y coordinate system rotated an angle, θ, with respect to the principal material axes system is given by

$$
\begin{aligned}
\epsilon_x &= \overline{S}_{11}\sigma_x + \overline{S}_{12}\sigma_y + \overline{S}_{16}\tau_{xy} \\
\epsilon_y &= \overline{S}_{21}\sigma_x + \overline{S}_{22}\sigma_y + \overline{S}_{26}\tau_{xy} \\
\gamma_{xy} &= \overline{S}_{61}\sigma_x + \overline{S}_{62}\sigma_y + \overline{S}_{66}\tau_{xy}
\end{aligned}
\qquad (2.54)
$$

The non-principal compliance terms are functions of the angle of rotation, θ, and the principal compliance terms, S_{ij}.

In addition, symmetry of the principal compliance matrix insures symmetry of the non-principal compliance matrix.

Table 2-1 Typical Material Properties*

	A	B	C	D	E	F
Property	Graphite/ Epoxy	Graphite/Epoxy AS/3501	GY 70/ Epoxy	E Glass/ Epoxy	Boron/Epoxy	Graphite/ Aluminum
E_1	20×10^6	21×10^6	42×10^6	6.2×10^6	30×10^6	18.0×10^6
E_2	2.1×10^6	1.4×10^6	1.0×10^6	1.7×10^6	2.1×10^6	3.6×10^6
ν_{12}	0.21	0.30	0.30	0.27	0.21	0.30
G_{12}	0.85×10^6	0.6×10^6	0.95×10^6	0.6×10^6	0.8×10^6	3.2×10^6
S_{11}	5.00×10^{-8}	4.76×10^{-8}	2.38×10^{-8}	1.61×10^{-7}	3.33×10^{-8}	5.56×10^{-8}
S_{12}	-1.05×10^{-8}	-1.43×10^{-8}	-7.14×10^{-9}	-4.35×10^{-8}	-7.00×10^{-9}	-1.67×10^{-8}
S_{22}	4.76×10^{-7}	7.14×10^{-7}	1.00×10^{-7}	5.88×10^{-7}	4.76×10^{-6}	2.78×10^{-7}
S_{66}	1.18×10^{-6}	1.67×10^{-6}	1.05×10^{-6}	1.67×10^{-6}	1.25×10^{-6}	3.13×10^{-6}
Q_{11}	2.01×10^7	2.11×10^7	4.29×10^7	6.33×10^8	3.07×10^7	1.83×10^7
Q_{12}	4.43×10^5	4.23×10^5	3.07×10^6	4.68×10^5	4.42×10^5	1.10×10^6
Q_{22}	2.11×10^6	1.40×10^6	1.02×10^7	1.73×10^6	2.11×10^6	3.67×10^6
Q_{66}	9.50×10^5	0.600×10^6	9.50×10^5	6.00×10^5	8.00×10^5	3.20×10^6

Footnote: *E_i, G_{12}, and Q_{ij} in psi, and S_{ij} in 1/psi. For SI units, 1 psi = 6.8948×10^3 Pa.

$$\begin{aligned}
\overline{S}_{11} &= m^4 S_{11} + m^2 n^2 (2S_{12} + S_{66}) + n^4 S_{22} \\
\overline{S}_{21} = \overline{S}_{12} &= m^2 n^2 (S_{11} + S_{22} - S_{66}) + S_{12}(m^4 + n^4) \\
\overline{S}_{22} &= n^4 S_{11} + m^2 n^2 (2S_{12} + S_{66}) + m^4 S_{22} \\
\overline{S}_{61} = \overline{S}_{16} &= 2m^3 n (S_{11} - S_{12}) + 2mn^3 (S_{12} - S_{22}) - mn(m^2 - n^2) S_{66} \\
\overline{S}_{62} = \overline{S}_{26} &= 2mn^3 (S_{11} - S_{12}) + 2m^3 n (S_{12} - S_{22}) + mn(m^2 - n^2) S_{66} \\
\overline{S}_{66} &= 4m^2 n^2 (S_{11} - S_{12}) - 4m^2 n^2 (S_{12} - S_{22}) + (m^2 - n^2)^2 S_{66}
\end{aligned} \tag{2.55}$$

In the same manner as the compliance relation, the stiffness relation may be established in an arbitrary coordinate system.

$$\begin{aligned}
\sigma_x &= \overline{Q}_{11}\epsilon_x + \overline{Q}_{12}\epsilon_y + \overline{Q}_{16}\gamma_{xy} \\
\sigma_y &= \overline{Q}_{21}\epsilon_x + \overline{Q}_{22}\epsilon_y + \overline{Q}_{26}\gamma_{xy} \\
\tau_{xy} &= \overline{Q}_{61}\epsilon_x + \overline{Q}_{62}\epsilon_y + \overline{Q}_{66}\gamma_{xy}
\end{aligned} \tag{2.56}$$

where the terms of the non-principal stiffness matrix are:

$$\begin{aligned}
\overline{Q}_{11} &= m^4 Q_{11} + 2m^2 n^2 (Q_{12} + 2Q_{66}) + n^4 Q_{22} \\
\overline{Q}_{21} = \overline{Q}_{12} &= m^2 n^2 (Q_{11} + Q_{22} - 4Q_{66}) + (m^4 + n^4) Q_{12} \\
\overline{Q}_{22} &= n^4 Q_{11} + 2m^2 n^2 (Q_{12} + 2Q_{66}) + m^4 Q_{22} \\
\overline{Q}_{61} = \overline{Q}_{16} &= m^3 n (Q_{11} - Q_{12}) + mn^3 (Q_{12} - Q_{22}) - 2mn(m^2 - n^2) Q_{66} \\
\overline{Q}_{62} = \overline{Q}_{26} &= mn^3 (Q_{11} - Q_{12}) + m^3 n (Q_{12} - Q_{22}) + 2mn(m^2 - n^2) Q_{66} \\
\overline{Q}_{66} &= m^2 n^2 (Q_{11} + Q_{22} - 2Q_{12} - 2Q_{66}) + (m^4 + n^4) Q_{66}
\end{aligned} \tag{2.57}$$

Typical material properties are shown in Table 2-1, and stiffness transformations are shown in Table 2-2 for graphite/epoxy.

2.4.5 Shear-Coupling Phenomena

It was pointed out in Sections 2.4.1 and 2.4.3 that the shear and normal components of stress and strain are uncoupled in the principal material coordinate systems. That is normal stresses produce only

Table 2-2 Transformed Stiffness Matrix*
$E_1 = 20\times10^6$ $E_2 = 2.1\times10^6$ $\nu_{12} = .21$ $G_{12} = .85\times10^6$

θ	$\overline{Q}_{11}$	$\overline{Q}_{12}$	$\overline{Q}_{16}$	$\overline{Q}_{22}$	$\overline{Q}_{26}$	$\overline{Q}_{66}$
0	2.009×10^7	4.431×10^5	0.	2.110×10^6	0.	8.500×10^5
45	6.622×10^6	4.922×10^6	4.496×10^6	6.622×10^6	4.496×10^6	5.329×10^6
90	2.110×10^6	4.431×10^5	0.	2.009×10^7	0.	8.500×10^5

*E_1, G_{12}, and $\overline{Q}_{ij}$ in psi. For SI units, 1 psi = 6.8948×10^3 Pa

normal strains and shear stress produces only shear strain. However, eqs (2.54) and (2.56) show that in an arbitrary coordinate system shear and normal response is coupled through the shear coupling compliance and stiffness terms $\overline{S}_{16}$, $\overline{S}_{26}$, $\overline{Q}_{16}$ and $\overline{Q}_{26}$. As a quantitative measure of this interaction the shear coupling ratios have been defined

$$\eta_{xy} = \frac{\gamma_{xy}}{\epsilon_x} \quad \text{when} \quad \sigma_y = \tau_{xy} = 0 \tag{2.58}$$

$$\eta_{yx} = \frac{\gamma_{xy}}{\epsilon_y} \quad \text{when} \quad \sigma_x = \tau_{xy} = 0 \tag{2.59}$$

If eqs (2.54) and (2.58-59) are combined the shear coupling ratio can be expressed as a function of the compliance terms

$$\eta_{xy} = \overline{S}_{16}/\overline{S}_{11} \tag{2.60}$$

$$\eta_{yx} = \overline{S}_{26}/\overline{S}_{22} \tag{2.61}$$

2.4.6 Thermoelastic Stress–Strain Relation

When an orthotropic material is subjected to a change in temperature the stress–strain relations must be modified to account for the free thermal strains. Equation (2.44) may be rewritten as follows:

$$\begin{aligned} \epsilon_1 - \alpha_1\Delta T &= \frac{\sigma_1}{E_1} - \frac{\nu_{21}\sigma_2}{E_2} + (0)\tau_{12} \\ \epsilon_2 - \alpha_2\Delta T &= \frac{-\nu_{12}\sigma_1}{E_1} + \frac{\sigma_2}{E_2} + (0)\tau_{12} \\ \epsilon_{12} - (0)\Delta T &= (0)\sigma_1 + (0)\sigma_2 + \frac{\tau_{12}}{G_{12}} \end{aligned} \tag{2.62}$$

where α_1 and α_2 are the coefficients of thermal expansion in the principal material coordinate system x_1,x_2, respectively, and T denotes the local temperature distribution. Equation (2.62) may be expressed in matrix form as before

$$\epsilon_i - \alpha_i \Delta T = S_{ij}\sigma_j \qquad (i,j = 1,2,6) \tag{2.63}$$

$$\sigma_i = Q_{ij}(\epsilon_j - \alpha_j \Delta T) \qquad (i,j = 1,2,6) \tag{2.64}$$

When it is desired to determine the thermoelastic stress–strain relation in an arbitrary coordinate system, it is necessary to determine the coefficients of thermal expansion in that coordinate system.

$$\begin{aligned} \alpha_x &= m^2\alpha_1 + n^2\alpha_2 \\ \alpha_y &= n^2\alpha_1 + m^2\alpha_2 \\ \alpha_{xy} &= 2mn\alpha_1 - 2mn\alpha_2 \end{aligned} \tag{2.65}$$

It should be noted that although α_6 is zero, α_{xy} vanishes in general only at $\theta = 0$ deg, 90 deg. This is illustrated in Fig. 2-4 where α_x/α_2 and α_{xy}/α_2 are shown as a function of θ for the case where α_1 vanishes. In addition, it is important to observe that α_{xy} is a shear coefficient of expansion. When $\alpha_{xy} \neq 0$ a change in temperature will produce a shear strain in the x-y coordinate system. The thermoelastic compliance and stiffness relations can now be determined for an arbitrary coordinate system.

$$\begin{aligned} \epsilon_x - \alpha_x \Delta T &= \overline{S}_{11}\sigma_x + \overline{S}_{12}\sigma_y + \overline{S}_{16}\tau_{xy} \\ \epsilon_y - \alpha_y \Delta T &= \overline{S}_{12}\sigma_x + \overline{S}_{22}\sigma_y + \overline{S}_{26}\tau_{xy} \\ \gamma_{xy} - \alpha_{xy} \Delta T &= \overline{S}_{16}\sigma_x + \overline{S}_{26}\sigma_y + \overline{S}_{66}\tau_{xy} \end{aligned} \tag{2.66}$$

and

$$\begin{aligned} \sigma_x &= \overline{Q}_{11}(\epsilon_x - \alpha_x \Delta T) + \overline{Q}_{12}(\epsilon_y - \alpha_y \Delta T) + \overline{Q}_{16}(\gamma_{xy} - \alpha_{xy} \Delta T) \\ \sigma_y &= \overline{Q}_{12}(\epsilon_x - \alpha_x \Delta T) + \overline{Q}_{22}(\epsilon_y - \alpha_y \Delta T) + \overline{Q}_{26}(\gamma_{xy} - \alpha_{xy} \Delta T) \\ \tau_{xy} &= \overline{Q}_{16}(\epsilon_x - \alpha_x \Delta T) + \overline{Q}_{26}(\epsilon_y - \alpha_y \Delta T) + \overline{Q}_{66}(\gamma_{xy} - \alpha_{xy} \Delta T) \end{aligned} \tag{2.67}$$

2.4.7 Hygroscopic - Thermal Analogy

The hygroscopic nature of polymeric composites results in free-hygroscopic strains when the moisture concentration of the material is changed. The hygroscopic response is totally analogous to the thermal response of the material. Hence, it is necessary to define two new material properties β_1 and β_2—the coefficients of hygroscopic expansion, respectively. Equations (2.63, 64) may now be rewritten for hygroscopic response.

$$\epsilon_i - \beta_i \Delta C = S_{ij}\sigma_j \qquad (i,j = 1,2,6) \tag{2.68}$$

$$\sigma_i = Q_{ij}(\epsilon_j - \beta_j \Delta C) \qquad (i,j = 1,2,6) \quad (2.69$$

and

$$\begin{aligned} \epsilon_x - \beta_x \Delta C &= \overline{S_{11}}\sigma_x + \overline{S_{12}}\sigma_y + \overline{S_{16}}\tau_{xy} \\ \epsilon_y - \beta_y \Delta C &= \overline{S_{12}}\sigma_x + \overline{S_{22}}\sigma_y + \overline{S_{26}}\tau_{xy} \\ \gamma_{xy} - \beta_{xy} \Delta C &= \overline{S_{16}}\sigma_x + \overline{S_{26}}\sigma_y + \overline{S_{66}}\tau_{xy} \end{aligned} \tag{2.70}$$

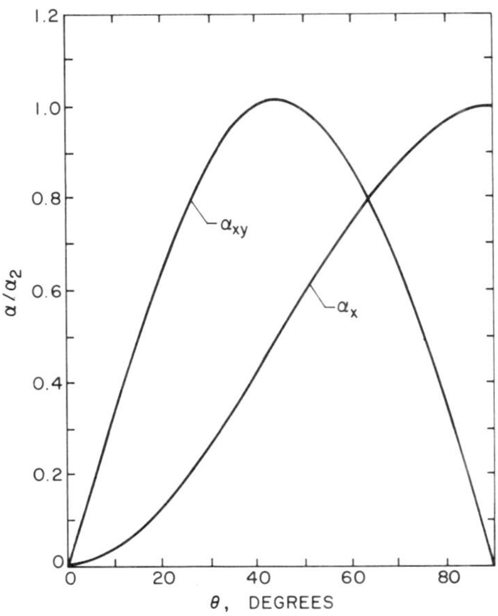

Fig. 2-4—Variation of unidirectional thermal expansion coefficients with orientation, $\alpha_1 = 0$

where

$$\beta_x = m^2\beta_1 + n^2\beta_2$$

$$\beta_y = n^2\beta_1 + m^2\beta_2$$

$$\beta_{xy} = 2mn\beta_1 - 2mn\beta_2 \; . \tag{2.71}$$

When $\beta_1 = 0$, eq (2.71) yields exactly the same results for β_x and β_{xy} as illustrated in Fig. 2-4 for thermal expansion coefficients.

2.5 MECHANICAL PROPERTIES OF THE LAMINATE

Composite laminates are constructed of an arbitrary number of orthotropic laminae with planes of elastic symmetry in the plane of the laminate. Laminate deformations are generally considered small with respect to the laminate thickness. For thin laminates, strains vary linearly through the thickness and interlaminar deformations may be considered small at interior regions. Although the laminate may consist of lamina of several materials, the behavior of each laminae must obey the linear stress–strain relations given in Section 2.4.

2.5.1 Laminate Stiffness Relation

In order to describe the behavior of the laminate it is necessary to define the laminate force and moment resultants. If the coordinate system of the laminate is *x-y*, then the following definitions can be made:

FORCE RESULTANTS:

$$N_x = \int_{-h/2}^{h/2} \sigma_x dx$$

$$N_y = \int_{-h/2}^{h/2} \sigma_y dz \tag{2.72}$$

$$N_{xy} = \int_{-h/2}^{h/2} \tau_{xy} dz$$

MOMENT RESULTANTS:

$$M_x = \int_{-h/2}^{h/2} \sigma_x z dz$$

$$M_y = \int_{-h/2}^{h/2} \sigma_y z dz \tag{2.73}$$

$$M_{xy} = \int_{-h/2}^{h/2} \tau_{xy} z dz$$

where h is the laminate thickness and z is the normal coordinate measured from the laminate mid-surface.

Since laminate strains are assumed in classical theory to vary linearly through the laminate thickness, it is possible to express the strain at any point through the laminate thickness as a function of laminate midplane strains, ϵ^o and curvatures, κ.

$$\begin{aligned} \epsilon_x &= \epsilon_x^o + z\kappa_x \\ \epsilon_y &= \epsilon_y^o + z\kappa_y \\ \gamma_{xy} &= \gamma_{xy}^o + z\kappa_{xy} \end{aligned} \tag{2.74}$$

Finally, it is possible to express the force and moment resultants in terms of the midplane strains and curvatures of the laminate.[7-9]

$$\begin{aligned} N_x &= A_{11}\epsilon_x^o + A_{12}\epsilon_y^o + A_{16}\gamma_{xy}^o + B_{11}\kappa_x + B_{12}\kappa_y + B_{16}\kappa_{xy} \\ N_y &= A_{12}\epsilon_x^o + A_{22}\epsilon_y^o + A_{26}\gamma_{xy}^o + B_{12}\kappa_x + B_{22}\kappa_y + B_{26}\kappa_{xy} \\ N_{xy} &= A_{16}\epsilon_x^o + A_{26}\epsilon_y^o + A_{66}\gamma_{xy}^o + B_{16}\kappa_x + B_{26}\kappa_y + B_{66}\kappa_{xy} \\ M_x &= B_{11}\epsilon_x^o + B_{12}\epsilon_y^o + B_{16}\gamma_{xy}^o + D_{11}\kappa_x + D_{12}\kappa_y + D_{16}\kappa_{xy} \\ M_y &= B_{12}\epsilon_x^o + B_{22}\epsilon_y^o + B_{26}\gamma_{xy}^o + D_{12}\kappa_x + D_{22}\kappa_y + D_{26}\kappa_{xy} \\ M_{xy} &= B_{16}\epsilon_x^o + B_{26}\epsilon_y^o + B_{66}\gamma_{xy}^o + D_{16}\kappa_x + D_{26}\kappa_y + D_{66}\kappa_{xy} \end{aligned} \tag{2.75}$$

where the terms of the laminate stiffness matrix are defined in terms of the laminae properties, $\overline{Q}_{ij}$, and h_k, the z coordinates of the kth interface as shown in Fig. 2-5. It should be noted that h_k is negative below the mid-plane and positive above the mid-plane.

$$\begin{aligned} A_{ij} &= \sum_{k=1}^{n} \overline{Q}_{ij}^k (h_k - h_{k-1}) \\ B_{ij} &= \frac{1}{2} \sum_{k=1}^{n} \overline{Q}_{ij}^k (h_k^2 - h_{k-1}^2) \\ D_{ij} &= \frac{1}{3} \sum_{k=1}^{n} \overline{Q}_{ij}^k (h_k^3 - h_{k-1}^3) \end{aligned} \tag{2.76}$$

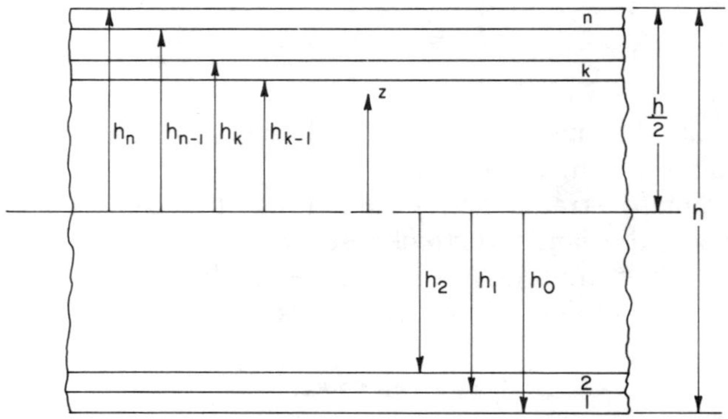

Fig. 2-5—Layer nomenclature for laminate

Several simplifications of eqs (2.75) are possible. First, consider the "balanced" laminate which contains an equal number of laminae of $+\theta$ deg and $-\theta$ deg fiber orientation. For such a laminate the shear coupling stiffness terms, A_{16} and A_{26}, vanish.

$$A_{16} = A_{26} = 0 \text{ (if laminate is balanced)}$$

Second, if the laminae within the laminate are positioned symmetrically with respect to the laminate mid-plane, then the coupling terms, B_{ij}, vanish.

$$B_{ij} = 0 \; (i,j = 1,2,6) \text{ (if laminate is symmetric)}$$

Thus the laminate stiffness relation for a balanced and symmetric laminate reduces to the following:

$$\begin{aligned}
N_x &= A_{11}\epsilon_x^o + A_{12}\epsilon_y^o \\
N_y &= A_{12}\epsilon_x^o + A_{22}\epsilon_y^o \\
N_{xy} &= A_{66}\gamma_{xy}^o \\
M_x &= D_{11}\kappa_x + D_{12}\kappa_y + D_{16}\kappa_{xy} \\
M_y &= D_{12}\kappa_x + D_{22}\kappa_y + D_{26}\kappa_{xy} \\
M_{xy} &= D_{16}\kappa_x + D_{26}\kappa_y + D_{66}\kappa_{xy}
\end{aligned} \tag{2.77}$$

Table 2-3 Laminate Properties: Material "A"

Term	$[0/45/-45]_s$	Laminate $[0/45/-45/45/-45/0]$	$[0/45/45]_s$
A_{11}	3.333×10^5 #/IN*	3.333×10^5 #/IN	3.333×10^5 #/IN
A_{12}	1.029×10^5 #/IN	1.029×10^5 #/IN	1.029×10^5 #/IN
A_{22}	1.535×10^5 #/IN	1.535×10^5 #/IN	1.535×10^5 #/IN
A_{66}	1.150×10^5 #/IN	1.150×10^5 #/IN	1.150×10^5 #/IN
A_{16}	0	0	8.992×10^4 #/IN
A_{26}	0	0	8.992×10^4 #/IN
B_{11}	0	0	0
B_{12}	0	0	0
B_{22}	0	0	0
B_{66}	0	0	0
B_{16}	0	2.248×10^2 #*	0
B_{26}	0	2.248×10^2 #	0
D_{11}	3.623×10^1 # IN*	3.623×10^1 # IN	3.623×10^1 # IN
D_{12}	3.983×10^0 # IN	3.983×10^0 # IN	3.983×10^0 # IN
D_{22}	7.755×10^0 # IN	7.755×10^0 # IN	7.755×10^0 # IN
D_{66}	4.899×10^0 # IN	4.899×10^0 # IN	4.899×10^0 # IN
D_{16}	2.248×10^0 # IN	0	2.997×10^0 # IN
D_{26}	2.248×10^0 # IN	0	2.997×10^0 # IN

**For SI units, 1 #/IN = 1.7513×10^2 N/m, 1 # = 4.4482 N, 1 # IN = 1.1298×10^{-1} N/m*

It is important to note that while the inplane shear and normal terms are uncoupled for the symmetric and balanced laminate, the twisting-bending terms (D_{16}, D_{26}) do not vanish.

Values of A_{ij}, B_{ij}, and D_{ij} are listed in Table 2-3 for laminates constructed of material "A" from Table 2-1. The thickness of each ply is 0.005 in.

2.5.2 *Effective Laminate Engineering Properties*

When dealing with composite laminates, it is often desirable to treat them as homogeneous and therefore possessing effective laminate engineering properties. For balanced, symmetric laminates it is possible to express the effective engineering properties as a function of the stiffness constants, A_{ij} and h the laminate thickness.

LONGITUDINAL YOUNG'S MODULUS:

$$E_x = (A_{11}A_{22} - A_{12}^2)/hA_{22} \tag{2.78}$$

TRANSVERSE YOUNG'S MODULUS:

$$E_y = (A_{11}A_{22} - A_{12}^2)/hA_{11} \tag{2.79}$$

LONGITUDINAL POISSON'S RATIO:

$$\nu_{xy} = A_{12}/A_{22} \tag{2.80}$$

TRANSVERSE POISSON'S RATIO:

$$\nu_{yx} = A_{12}/A_{11} \tag{2.81}$$

SHEAR MODULUS:

$$G_{xy} \doteq A_{66}/h \tag{2.82}$$

If the laminate is not symmetric, determination of effective laminate properties is more complex. The relation (2.75) must first be expressed in matrix form and inverted. This can be accomplished with contemporary matrix inversion schemes using digital computation facilities. In the inverted form, eq (2.75) becomes:

$$\begin{aligned}
\epsilon_x^o &= H_{11}N_x + H_{12}N_y + H_{13}N_{xy} + H_{14}M_x + H_{15}M_y + H_{16}M_{xy} \\
\epsilon_y^o &= H_{12}N_x + H_{22}N_y + H_{23}N_{xy} + H_{24}M_x + H_{25}M_y + H_{26}M_{xy} \\
\gamma_{xy}^o &= H_{13}N_x + H_{23}N_y + H_{33}N_{xy} + H_{34}M_x + H_{35}M_y + H_{36}M_{xy} \\
\varkappa_x &= H_{14}N_x + H_{24}N_y + H_{34}N_{xy} + H_{44}M_x + H_{45}M_y + H_{46}M_{xy} \\
\varkappa_y &= H_{15}N_x + H_{25}N_y + H_{35}N_{xy} + H_{45}M_x + H_{55}M_y + H_{56}M_{xy} \\
\varkappa_{xy} &= H_{16}N_x + H_{26}N_y + H_{36}N_{xy} + H_{46}M_x + H_{56}M_y + H_{66}M_{xy}
\end{aligned} \tag{2.83}$$

Then the effective elastic properties may be written in terms of the H_{ij}.

LONGITUDINAL YOUNG'S MODULUS:

$$E_x = (hH_{11})^{-1} \tag{2.84}$$

TRANSVERSE YOUNG'S MODULUS:

$$E_y = (hH_{22})^{-1} \tag{2.85}$$

Table 2-4 Laminate Effective Properties: Material "A"

Prop.	$[0/+45/-45]_s$	$[0/45/-45/45/-45/0]$	$[0/45/45]_s$
E_x	8.82×10^6 PSI*	8.78×10^6 PSI	8.34×10^6 PSI
E_y	4.06×10^6 PSI	3.89×10^6 PSI	2.64×10^6 PSI
ν_{xy}	0.670	0.646	0.392
ν_{yx}	0.309	0.286	0.124
G_{xy}	3.84×10^6 PSI	3.61×10^6 PSI	1.97×10^6 PSI

**For SI units, 1 psi = 6.8948×10^3 Pa*

LONGITUDINAL POISSON'S RATIO:

$$\nu_{xy} = -H_{12}/H_{11} \tag{2.86}$$

TRANSVERSE POISSON'S RATIO:

$$\nu_{yx} = -H_{12}/H_{22} \tag{2.87}$$

SHEAR MODULUS:

$$G_{xy} = (hH_{33})^{-1} \tag{2.88}$$

Effective laminate properties are listed in Table 2-4 for laminates constructed of material "A" from Table 2-1. It should be noted that those constants may be very difficult to measure experimentally for unsymmetric laminates due to bending-extensional coupling.

2.5.3 *Coefficients of Thermal Expansion*

If the laminae coefficients of thermal expansion and the laminate configuration are known, it is possible to develop relations for the effective laminate coefficients of thermal expansion. First, it is necessary to define the thermal force resultants in terms of the laminae stiffnesses, $\overline{Q}_{ij}^k$, and coefficients of thermal expansion, $\overline{\alpha}_j^k$,

THERMAL FORCE RESULTANTS:

$$\begin{aligned} N_x^T &= \sum_{k=1}^{n} (\overline{Q}_{11}^k \overline{\alpha}_1^k + \overline{Q}_{12} \overline{\alpha}_2^k + \overline{Q}_{16}^k \overline{\alpha}_6^k)(h_k - h_{k-1})\Delta T \\ N_y^T &= \sum_{k=1}^{n} (\overline{Q}_{12}^k \overline{\alpha}_1^k + \overline{Q}_{22}^k \overline{\alpha}_2^k + \overline{Q}_{26}^k \overline{\alpha}_6^k)(h_k - h_{k-1})\Delta T \\ N_{xy}^T &= \sum_{k=1}^{n} (\overline{Q}_{16}^k \overline{\alpha}_1^k + \overline{Q}_{26}^k \overline{\alpha}_2^k + \overline{Q}_{66}^k \overline{\alpha}_6^k)(h_k - h_{k-1})\Delta T \end{aligned} \tag{2.89}$$

where $\bar{\alpha}_1$, $\bar{\alpha}_2$, and $\bar{\alpha}_6$ are equivalent to α_x^k, α_y^k and α_{xy}^k for the kth lamina. The effective laminate coefficients of thermal expansion for a balanced-symmetric laminate may now be written in terms of eqs (2.89) and (2.77).

LONGITUDINAL COEFFICIENT OF THERMAL EXPANSION:

$$\alpha_x = \frac{A_{22}N_x^T - A_{12}N_y^T}{(A_{11}A_{22} - A_{12}^2)\Delta T} \tag{2.90}$$

TRANSVERSE COEFFICIENT OF THERMAL EXPANSION:

$$\alpha_y = \frac{A_{11}N_y^T - A_{12}N_x^T}{(A_{11}A_{22} - A_{12}^2)\Delta T} \tag{2.91}$$

SHEAR COEFFICIENT OF THERMAL EXPANSION:

$$\alpha_{xy} = N_{xy}^T/A_{66}\Delta T = 0. \tag{2.92}$$

Thus for the balanced-symmetric laminate, α_x and α_y are principal coefficients of thermal expansion. Laminate coefficient of thermal expansion for any orientation, θ, with respect to the laminate longitudinal axis may now be determined:

$$\alpha_\theta = m^2\alpha_x + n^2\alpha_y \tag{2.93}$$

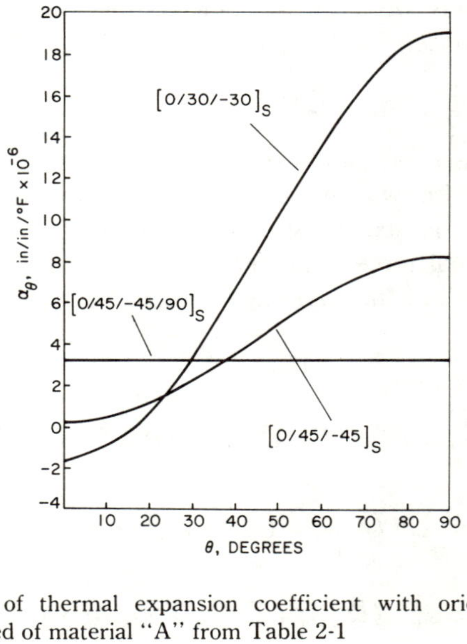

Fig. 2-6—Variation of thermal expansion coefficient with orientation for laminates constructed of material "A" from Table 2-1

Table 2-5 Thermal Loads: Material "A"

Laminate	$N_x^T/\Delta T$*	$N_y^T/\Delta T$	$N_{xy}^T/\Delta T$
$[0/45/-45]_s$	0.929	1.302	0.00
$[0/45/-45/45/-45/0]$	0.929	1.302	0.00
$[0/45/45]_s$	0.929	1.302	−0.373
Laminate	$M_x^T/\Delta T$**	$M_y^T/\Delta T$	$M_{xy}^T/\Delta T$
$[0/45/-45]_s$	0.00	0.00	0.00
$[0/45/-45/45/-45/0]$	0.00	0.00	-9.33×10^{-4}
$[0/45/45]_s$	0.00	0.00	0.00

$\alpha_1 = 0.34 \times 10^{-6}$ IN/IN/°F
$\alpha_2 = 26.4 \times 10^{-6}$ IN/IN/°F

#/IN/°F. In SI units, 1 #/IN/°F = 9.7294 N/m/°k
***#/°F. In SI units, 1 #/°F = 2.4711 N/°k*

Table 2-6 The Laminate Effective Thermal Coefficients for Material "A"

Laminate	$\bar{\alpha}_x$ (IN/IN/°F)*	$\bar{\alpha}_y$ (IN/IN/°F)
$[0/45/-45]_s$	2.14×10^{-7}	8.34×10^{-6}
$[0/45/-45/90]_s$	3.22×10^{-6}	3.22×10^{-6}
$[0/30/-30]_s$	-1.75×10^{-6}	1.895×10^{-5}

$\alpha_1 = 0.34\ 10^{-6}$ IN/IN/°F
$\alpha_2 = 26.4\ 10^{-6}$ IN/IN/°F

**In SI units, 1 m/m/°K = 1.8 IN/IN°F*

where, as before, $m = \cos\theta$ and $n = \sin\theta$. In general the shear coefficient of expansion, $\alpha_{\theta x}$, will not vanish.

$$\alpha_{\theta x} = 2mn(\alpha_x - \alpha_y) \tag{2.94}$$

Thermal loads are listed in Table 2-5 and effective laminate thermal coefficients in Table 2-6 for laminates constructed of material "A" from Table 2-1. The numbers in Table 2-5 are based on 0.005 in ply thickness. Equation (2.93) is illustrated in Fig. 2-6 for laminates constructed of the same material.

2.5.4 Coefficients of Hygroscopic Expansion

Effective laminate coefficients of hygroscopic expansion can be developed from the thermal analogy discussed in section 2.4.7. Thus, eq (2.89) becomes

$$
\begin{aligned}
N_x^H &= \sum_{k=1}^{n} (\overline{Q}_{11}^k \overline{\beta}_1^k + \overline{Q}_{12}^k \overline{\beta}_2^k + \overline{Q}_{16}^k \overline{\beta}_6^k)(h_k - h_{k-1})\,\Delta C \\
N_y^H &= \sum_{k=1}^{n} (\overline{Q}_{12}^k \overline{\beta}_1^k + \overline{Q}_{22}^k \overline{\beta}_2^k + \overline{Q}_{26}^k \overline{\beta}_6^k)(h_k - h_{k-1})\,\Delta C \qquad (2.95) \\
N_{xy}^H &= \sum_{k=1}^{n} (\overline{Q}_{16}^k \overline{\beta}_1^k + \overline{Q}_{26}^k \overline{\beta}_2^k + \overline{Q}_{66}^k \overline{\beta}_6^k)(h_k - h_{k-1})\,\Delta C
\end{aligned}
$$

where N_x^H, N_y^H, and N_{xy}^H are hygroscopic force resultants; and $\overline{\beta}_1$, $\overline{\beta}_2$, and $\overline{\beta}_6$ are analogous to $\overline{\alpha}_1$, $\overline{\alpha}_2$ and $\overline{\alpha}_6$, respectively. Equations (2.90-2.92) for effective thermal expansion coefficients of balanced-symmetric laminates now become

$$
\begin{aligned}
\beta_x &= \frac{A_{22}N_x^H - A_{12}N_y^H}{(A_{11}A_{22} - A_{12}^2)\Delta C} \\
\beta_y &= \frac{A_{11}N_y^H - A_{12}N_x^H}{(A_{11}A_{22} - A_{12}^2)\Delta C} \qquad (2.96) \\
\beta_{xy} &= N_{xy}^H / A_{66}\Delta c = 0
\end{aligned}
$$

where β_x and β_y are the coefficients of hygroscopic expansion in the longitudinal and transverse directions, respectively, and β_{xy} is the hygroscopic shear expansion coefficient. The transformations for laminate coefficients of hygroscopic expansion are analogous to eqs (2.93) and (2.94), i.e.,

$$
\begin{aligned}
\beta_\theta &= m^2\beta_x + n^2\beta_y \\
\beta_{\theta x} &= 2mn(\beta_x - \beta_y)
\end{aligned}
\qquad (2.97)
$$

2.5.5 Combined Expansional Strains

In many practical applications both moisture and temperature will be present. For such a case, it is necessary to account for the combined effect of thermal expansion and hygroscopic expansion utilizing generalized expansional strains, eq (2.19). In order to develop effective laminate expansional strains, it is necessary to define expansional force resultants

$$N_x^E = N_x^T + N_x^H$$

$$N_y^E = N_y^T + N_y^H \tag{2.98}$$

$$N_{xy}^E = N_{xy}^T + N_{xy}^H$$

The effective expansional strains for a balanced-symmetric laminate are of the form

$$\epsilon_x^E = \frac{A_{22}N_x^E - A_{12}N_y^E}{A_{11}A_{22} - A_{12}^2}$$

$$\epsilon_y^E = \frac{A_{11}N_y^E - A_{12}N_x^E}{A_{11}A_{22} - A_{12}^2} \tag{2.99}$$

$$\epsilon_{xy}^E = N_{xy}^E/A_{66} = 0$$

where ϵ_x^E and ϵ_y^E are the longitudinal and transverse expansional strains, respectively, and ϵ_{xy}^E is the expansional shear strain. The transformations for laminate expansional strains are of the same form as eqs (2.93) and (2.94), i.e.,

$$\epsilon_\theta^E = m^2\epsilon_x^E + n^2\epsilon_y^E$$

$$\epsilon_{\theta x}^E = 2mn(\epsilon_x^E - \epsilon_y^E) \tag{2.100}$$

2.5.6 Variation of Laminate Properties with Orientation

Effective elastic properties are often desired for any orientation, θ, with respect to the longitudinal material axis of a balanced, symmetric laminate. Integration of eqs (2.57) through the laminate thickness yields

$$\overline{A}_{11} = m^4A_{11} + 2m^2n^2(A_{12} + 2A_{66}) + n^4A_{22}$$

$$\overline{A}_{12} = \overline{A}_{21} = m^2n^2(A_{11} + A_{22} - 4A_{66}) + (m^4 + n^4)A_{12}$$

$$\overline{A}_{22} = n^4A_{11} + 2m^2n^2(A_{12} + 2A_{66}) + m^4A_{22} \tag{2.101}$$

$$\overline{A}_{16} = \overline{A}_{61} = m^3n(A_{11} - A_{12}) + mn^3(A_{12} - A_{22}) - 2mn(m^2 - n^2)A_{66}$$

$$\overline{A}_{26} = \overline{A}_{62} = mn^3(A_{11} - A_{12}) + m^3n(A_{12} - A_{22}) + 2mn(m^2 - n^2)A_{66}$$

$$\overline{A}_{66} = m^2n^2(A_{11} + A_{22} - 2A_{12} - 2A_{66}) + (m^4 + n^4)A_{66}$$

where $\overline{A}_{ij}$ are the transformed inplane stiffnesses. If we denote the inverted values of $\overline{A}_{ij}$ by $\overline{L}_{ij}$, then the transformed elastic properties take the form

$$\begin{aligned} E_{x'} &= (h\overline{L}_{11})^{-1} \\ E_{y'} &= (h\overline{L}_{22})^{-1} \\ \nu_{x'y'} &= -\overline{L}_{12}/\overline{L}_{11} \\ \nu_{y'x'} &= -\overline{L}_{12}/\overline{L}_{22} \\ G_{x'y'} &= (h\overline{L}_{66})^{-1} \end{aligned} \tag{2.102}$$

where x',y' are the rotated x,y axes, respectively.

Consider the special case where $A_{11} = A_{22}$ and $(A_{11} - A_{12} - 2A_{66}) = 0$. A cursory examination of eq (2.101) reveals that

$$\begin{aligned} \overline{A}_{11} = \overline{A}_{22} = A_{11}, \quad & \overline{A}_{12} = \overline{A}_{21} = A_{12}, \\ \overline{A}_{16} = \overline{A}_{26} = 0, \quad & \overline{A}_{66} = A_{66} \end{aligned} \tag{2.103}$$

Thus, the inplane stiffnesses are independent of θ, yielding an isotropic material, in this plane, i.e.,

$$\begin{aligned} E_{x'} = E_{y'} = E_x, \quad & \nu_{x'y'} = \nu_{y'x'} = \nu_{xy}, \\ G_{x'y'} = G_{xy} = & \frac{E_x}{2(1 + \nu_{xy})} \end{aligned} \tag{2.104}$$

It can be shown that any balanced, symmetric laminate constructed of $2N$ plies oriented at $\theta = \pi/N$ radians, where $N \geq 3$, yields an inplane stiffness matrix which is isotropic. Such a laminate is referred to as *quasi-isotropic.* The term quasi-isotropic is utilized because the bending stiffness matrix, D_{ij}, will not be isotropic in most cases. As a result the laminate is isotropic for inplane loading only.

It can be easily seen from eq (2.100) that the effective expansional strains ϵ_θ^E are independent of θ whenever $\epsilon_y = \epsilon_x$. Thus, in the case of thermal expansion and hygroscopic expansion, effective isotropic properties are obtained any time a balanced, symmetric laminate has equal properties in the two principal material directions.

2.6 STRENGTH OF THE COMPOSITE LAMINATE

The multidirectional composite laminate consists of laminae of various fiber orientations. In this form the stiffness of each lamina may differ significantly from adjacent laminae. Since the strain components in thin laminates vary linearly through the laminate thickness, discontinuities in the inplane stress components will occur at laminae interfaces. In addition, the initiation of laminate failure may occur at an interior lamina wherein the magnitude of the stress components as compared to the intrinsic strength properties is maximum. Hence, it is imperative that the detailed state of stress be established for each layer. Fortunately, it is possible to establish subsurface laminae stresses and strains while measuring only surface laminate strains, for linear elastic behavior.

2.6.1 Lamina Mechanical Stresses

If laminate upper and lower surface strains are measured, it is possible to establish the state of strain within each lamina. Let the surface strains be denoted as follows:

$\epsilon_x^u, \epsilon_y^u, \epsilon_{xy}^u$ upper surface strain components at $z = h/2$

$\epsilon_x^\ell, \epsilon_y^\ell, \epsilon_{xy}^\ell$ lower surface strain components at $z = -h/2$

The laminate midplane strains and curvatures may now be expressed as a function of the upper and lower surface strain components.

$$
\begin{aligned}
\epsilon_x^o &= (\epsilon_x^u + \epsilon_x^\ell)/2 \\
\epsilon_y^o &= (\epsilon_y^u + \epsilon_y^\ell)/2 \\
\gamma_{xy}^o &= (\gamma_{xy}^u + \gamma_{xy}^\ell)/2
\end{aligned}
\tag{2.105}
$$

$$
\begin{aligned}
\kappa_x &= (\epsilon_x^u - \epsilon_x^\ell)/h \\
\kappa_y &= (\epsilon_y^u - \epsilon_y^\ell)/h \\
\kappa_{xy} &= (\gamma_{xy}^u - \gamma_{xy}^\ell)/h
\end{aligned}
\tag{2.106}
$$

Having established the laminate mid-plane strains and curvatures, one can determine the strain at any position throughout the thickness of the laminate. Combining eqs (2.105-6) and (2.74) yields

$$
\begin{aligned}
\epsilon_x &= (\epsilon_x^u + \epsilon_x^\ell)/2 + (\epsilon_x^u - \epsilon_x^\ell)z/h \\
\epsilon_y &= (\epsilon_y^u + \epsilon_y^\ell)/2 + (\epsilon_y^u - \epsilon_y^\ell)z/h \\
\gamma_{xy} &= (\gamma_{xy}^u + \gamma_{xy}^\ell)/2 + (\gamma_{xy}^u - \gamma_{xy}^\ell)z/h
\end{aligned}
\tag{2.107}
$$

Now that the strains throughout the laminate are known, it is possible to establish the stress components within the kth lamina.

$$\begin{aligned}\sigma_x^k = {} & \overline{Q}_{11}^k[(\epsilon_x^u + \epsilon_x^\ell)/2 + (\epsilon_x^u - \epsilon_x^\ell)z/h] \\ & + \overline{Q}_{12}^k[(\epsilon_y^u + \epsilon_y^\ell)/2 + (\epsilon_y^u - \epsilon_y^\ell)z/h] \\ & + \overline{Q}_{16}^k[(\gamma_{xy}^u + \gamma_{xy}^\ell)/2 + (\gamma_{xy}^u - \gamma_{xy}^\ell)z/h]\end{aligned} \tag{2.108}$$

$$\begin{aligned}\sigma_y^k = {} & \overline{Q}_{12}^k[(\epsilon_x^u + \epsilon_x^\ell)/2 + (\epsilon_x^u - \epsilon_x^\ell)z/h] \\ & + \overline{Q}_{22}^k[(\epsilon_y^u + \epsilon_y^\ell)/2 + (\epsilon_y^u - \epsilon_y^\ell)z/h] \\ & + \overline{Q}_{26}^k[(\gamma_{xy}^u + \gamma_{xy}^\ell)/2 + (\gamma_{xy}^u - \gamma_{xy}^\ell)z/h]\end{aligned} \tag{2.109}$$

$$\begin{aligned}\tau_{xy}^k = {} & \overline{Q}_{16}^k[(\epsilon_x^u + \epsilon_x^\ell)/2 + (\epsilon_x^u - \epsilon_x^\ell)z/h] \\ & + \overline{Q}_{26}^k[(\epsilon_y^u + \epsilon_y^\ell)/2 + (\epsilon_y^u - \epsilon_y^\ell)z/h] \\ & + \overline{Q}_{66}^k[(\gamma_{xy}^u + \gamma_{xy}^\ell)/2 + (\gamma_{xy}^u - \gamma_{xy}^\ell)z/h].\end{aligned} \tag{2.110}$$

If one establishes the position of the kth lamina as $z^k \leq z \leq z^{k+1}$, then it is desirable to establish the maximum stresses within that region. The

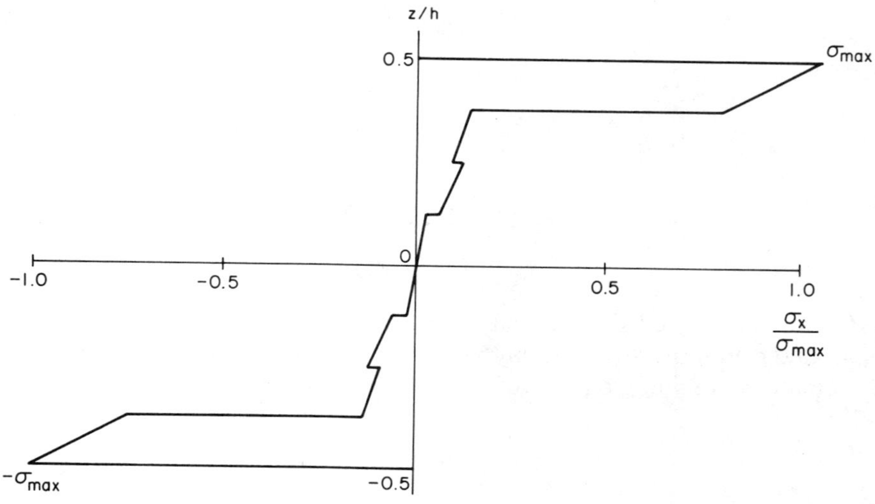

Fig. 2-7—Bending stress distribution, σ_x, for $[0/45-45/90]_s$ laminate constructed from material "A" in Table 2-1

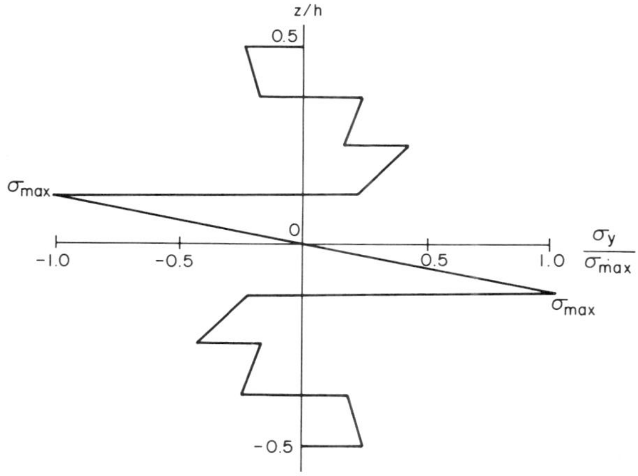

Fig. 2-8—Bending stress distribution, σ_y, for $[0/45-45/90]_s$ laminate constructed from material "A" in Table 2-1

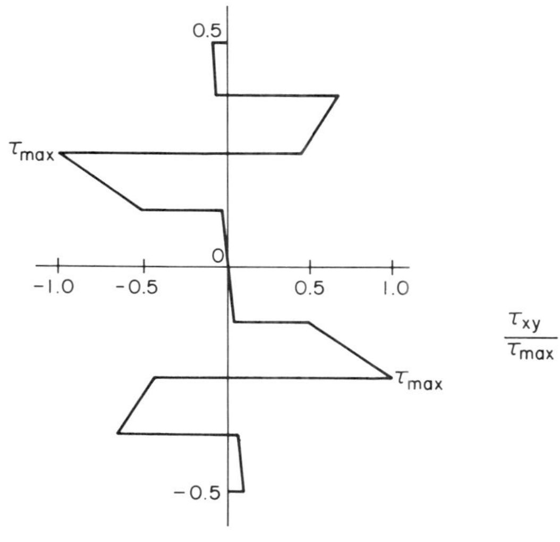

Fig. 2-9—Bending shear stress distribution, τ_{xy}, for $[0/45-45/90]_s$ laminate constructed from material "A" in Table 2-1

maximum normal stresses will occur at one of the interfaces $z = z^k$ or $z = z^{k+1}$. Bending stress distributions are shown in Figs. 2-7, 8, 9 for a $[0/45/-45/90]_s$ laminate constructed of material "A" in Table 2-1.

2.6.2 Lamina Thermal Residual Stresses and Strains

Laminae coefficients of thermal expansion are a function of fiber orientation. Since polymerization of polymeric composites is generally achieved by thermal processing, multidirectional laminates will possess thermal residual stresses in the processed condition. Calculation of thermal residual stresses and strains is possible by establishing the effective thermal force (2.89) and moment resultants.

THERMAL MOMENT RESULTANTS

$$M_x^T = \frac{1}{2}\sum_{k=1}^{n}(\overline{Q}_{11}^k\overline{\alpha}_1^k + \overline{Q}_{12}^k\overline{\alpha}_2^k + \overline{Q}_{16}^k\overline{\alpha}_6^k)(h_k^2 - h_{k-1}^2)\Delta T$$

$$M_y^T = \frac{1}{2}\sum_{k=1}^{n}(\overline{Q}_{12}^k\overline{\alpha}_1^k + \overline{Q}_{22}^k\overline{\alpha}_2^k + \overline{Q}_{26}^k\overline{\alpha}_6^k)(h_k^2 - h_{k-1}^2)\Delta T \qquad (2.111)$$

$$M_{xy}^T = \frac{1}{2}\sum_{k=1}^{n}(\overline{Q}_{16}^k\overline{\alpha}_1^k + \overline{Q}_{26}^k\overline{\alpha}_2^k + \overline{Q}_{66}^k\overline{\alpha}_6^k)(h_k^2 - h_{k-1}^2)\Delta T.$$

Combining eqs (2.89), (2.111) and (2.83), one can establish the laminate midplane strains and curvatures which result from the processing temperature, ΔT. In general ΔT is a negative number. Thus, for example, the curvature term κ_x is found by writing out the fourth equation in (2.83) for the present thermal bending case:

$$\kappa_x = H_{14}N_x^T + H_{24}N_y^T + H_{34}N_{xy}^T + H_{44}M_x^T + H_{45}M_y^T + H_{46}M_{xy}^T$$

Note should be made that for symmetric laminates $H_{14} = H_{24} = H_{34} = H_{15} = H_{25} = H_{35} = H_{16} = H_{26} = H_{36} = 0$. Thus, the as-processed laminate should exhibit no curvature as a result of the thermal processing. However, processing of unsymmetric laminates generally results in laminates possessing bending and twisting curvatures: κ_x, κ_y and κ_{xy}.

For symmetric and balanced laminates the thermal residual stresses may be determined by first calculating the resulting laminate strains.

$$\epsilon_x^o = \frac{A_{22}N_x^T - A_{12}N_y^T}{A_{11}A_{22} - A_{12}^2}$$

$$\epsilon_y^o = \frac{A_{11}N_y^T - A_{12}N_x^T}{A_{11}A_{22} - A_{12}^2} \qquad (2.112)$$

$$\gamma_{xy}^o = \kappa_x = \kappa_y = \kappa_{xy} = 0$$

Then the thermal residual stresses in the kth lamina may be calculated by combining eqs (2.108) and (2.85) to yield

$$\sigma_x^k = [(\overline{Q}_{11}^k A_{22} - \overline{Q}_{12}^k A_{12})N_x^T + (\overline{Q}_{12}^k A_{11} - \overline{Q}_{11}^k A_{12})N_y^T]/(A_{11}A_{22} - A_{12}^2)$$

$$\sigma_y^k = [(\overline{Q}_{12}^k A_{22} - \overline{Q}_{22}^k A_{12})N_x^T + (\overline{Q}_{22}^k A_{11} - \overline{Q}_{11}^k A_{12})N_y^T]/(A_{11}A_{22} - A_{12}^2)$$

$$\tau_{xy}^k = [(\overline{Q}_{16}^k A_{22} - \overline{Q}_{26}^k A_{12})N_x^T + (\overline{Q}_{26}^k A_{11} - \overline{Q}_{26}^k A_{12})N_y^T]/(A_{11} - A_{22} - A_{12}^2) \quad (2.113)$$

The above thermal residual stresses must be considered when laminate strength analyses are performed. Residual stress distributions are shown in Fig. 2-10 for a $[0/45/-45/90]_s$ laminate constructed of material "A" in Table 2-1.

It should be noted that ΔT is not necessarily the difference between use temperature and the cure temperature, due to the viscoelastic nature of polymeric matrix materials. A more detailed discussion of residual stresses can be found in Refs. 10 and 11.

In actual service, residual stresses will be modified by the presence of moisture. For such cases residual stress calculations must be based

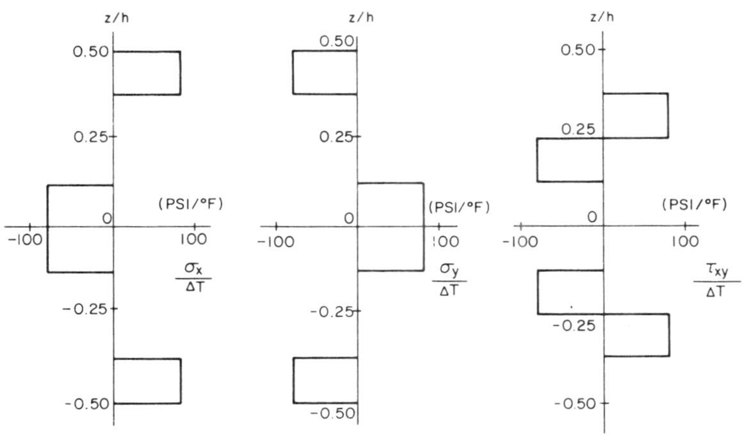

Fig. 2-10—Residual thermal stresses for $[0/45-45/90]_s$ laminate constructed of material "A" in Table 2-1

on generalized expansional strains, eq (2.19). These calculations are further complicated by the fact that the moisture distribution is rarely uniform in actual service. Residual stresses produced by moisture have been discussed by Hahn and Kim.[12] The presence of moisture in the matrix can also alter the residual stresses during fabrication.

2.6.3 Lamina Strength Models

Three models for lamina strength will be discussed in this section: maximum stress, maximum strain, and quadratic interaction criteria. The maximum stress model for failure may be stated as follows:

Maximum stress criterion

Failure results when any one of the stress components is equal to its corresponding intrinsic strength property. Expressed in equation form the maximum stress criterion is given by

$$\begin{aligned}
&\sigma_1 \geq X_1^T, \quad \sigma_1 > 0; \quad |\sigma_1| \geq X_1^c, \quad \sigma_1 < 0 \\
&\sigma_2 \geq X_2^T, \quad \sigma_2 > 0; \quad |\sigma_2| \geq X_2^c, \quad \sigma_2 < 0 \qquad (2.114) \\
&\tau_{12} \geq S_6^+, \quad \tau_{12} > 0; \quad |\tau_{12}| \geq S_6^-, \quad \tau_{12} < 0
\end{aligned}$$

where the intrinsic strengths are defined as follows:

X_1^T = ultimate tensile strength in fiber (1) direction
X_1^c = ultimate compressive strength in fiber (1) direction
X_2^T = ultimate tensile strength in transverse (2) direction
X_2^c = ultimate compressive strength in transverse (2) direction
S_6^+ = ultimate positive inplane (1-2) shear strength*
S_6^- = ultimate negative inplane (1-2) shear strength*

The maximum strain criterion is totally analogous to the maximum stress criterion.

Maximum strain criterion.

Failure results when any one of the strain components is equal to its corresponding intrinsic strength property. Expressed in equation form the maximum strain criterion is given as follows:

---

*$S_6^+ = S_6^-$ *for most composite systems.*

$$\epsilon_1 \geq e_1^T, \quad \epsilon_1 > 0; \quad |\epsilon_1| \geq e_1^c, \quad \epsilon_1 < 0$$

$$\epsilon_2 \geq e_2^T, \quad \epsilon_2 > 0; \quad |\epsilon_2| \geq e_2^c, \quad \epsilon_2 < 0 \tag{2.115}$$

$$\gamma_{12} \geq e_6^+, \quad \gamma_{12} > 0; \quad |\gamma_{12}| \geq e_6^-, \quad \gamma_{12} < 0$$

The intrinsic ultimate strains are defined as follows:

e_1^T = Ultimate tensile strain in fiber (1) direction
e_1^c = Ultimate compressive strain in fiber (1) direction
e_2^T = Ultimate tensile strain in transverse (2) direction
e_2^c = Ultimate compressive strain in transverse (2) direction
e_6^+ = Ultimate positive inplane (1-2) shear strain*
e_6^- = Ultimate negative inplane (1-2) shear strain*

Several forms of the quadratic interaction failure criterion may be found in the literature.[13-16] One of the most consistent of this class of failure models was developed recently by Tsai and Wu,[17] as a modification of a theory proposed by Goldenblat and Kopnov.[18] The quadratic interaction criterion accounts for interaction of stress components in determining strength in a biaxial stress field. When consideration is limited to specially orthotropic materials subjected to planar states of stress, the Tsai-Wu criterion may be expressed in equation form as follows:

$$F_1\sigma_1 + F_2\sigma_2 + F_6\sigma_6 + F_{11}\sigma_1^2 + 2F_{12}\sigma_1\sigma_2 + F_{22}\sigma_2^2 + F_{16}\sigma_1\sigma_6 + F_{26}\sigma_2\sigma_6 + F_{66}\sigma_6^2 = 1 \tag{2.116}$$

All of the constants of the quadratic failure criterion may be expressed in terms of the intrinsic ultimate strength properties of the materials, except F_{12}. If $S_6^+ = S_6^-$, then

$$F_1 = 1/X_1^T - 1/X_1^c \qquad F_{11} = 1/X_1^T X_1^c$$

$$F_2 = 1/X_2^T - 1/X_2^c \qquad F_{22} = 1/X_2^T X_2^c \tag{2.117}$$

$$F_6 = F_{16} = F_{26} = 0 \qquad F_{66} = 1/S_6^2$$

*$e_6^+ = e_6^-$ *for most composite systems.*

Narayanaswami and Adelman[19] were able to correlate eq (2.116) with experimental data by choosing F_{12} to be zero. Tsai, however, has recently suggested[20] the following relationship

$$F_{12} = -\frac{1}{2\sqrt{X_1^T X_2^T X_1^c X_1^c}} \tag{2.118}$$

If F_{12} is determined in this manner, it has been shown[20] that eq (2.116) becomes a generalization of the von Mises criterion for isotropic materials.

Now consider the case where

$$X_1^t = X_1^c = X_1, \quad X_2^T = X_2^c = X_2, \quad S_6^+ = S_6^- = S_6 \tag{2.119}$$

and

$$F_{12} = -\frac{1}{2X_1^2} \tag{2.120}$$

Equation (2.116) reduces to the Tsai-Hill criterion.[21]

$$\left(\frac{\sigma_1}{X_1}\right)^2 - \frac{\sigma_1\sigma_2}{X_1^2} + \left(\frac{\sigma_2}{X_2}\right)^2 + \left(\frac{\sigma_6}{S_6}\right)^2 = 1 \tag{2.121}$$

Thus, the Tsai-Hill criterion can be considered as a special case of the quadratic interaction criterion.

In order to demonstrate the importance of interaction of stress components in determining strength it is instructive to examine the tensile strength of coupons of arbitrary fiber orientation with respect to the loading axis, θ. The maximum stress criterion predicts failure in three distinct modes while the quadratic interaction criterion yields a continuous function of strength versus fiber orientation.

Maximum stress criterion

$$\begin{aligned} \sigma^{ult} &= 2X_1^T/(1 + \cos 2\theta) && \text{fiber tension failure} \\ \sigma^{ult} &= 2X_2^T/(1 - \cos 2\theta) && \text{transverse tension failure} \\ \sigma^{ult} &= 2S_6/\sin 2\theta && \text{inplane shear failure} \end{aligned} \tag{2.122}$$

Quadratic interaction criterion

$$\sigma^{ult} = \frac{-B + \sqrt{B^2 + 4A}}{2A} \quad * \tag{2.123}$$

where

$$A = \frac{F_{11}}{4}(1 + \cos 2\theta)^2 + \frac{F_{22}}{4}(1 - \cos 2\theta)^2 + \frac{F_{66}}{4}\sin^2 2\theta$$

$$B = \frac{F_1}{2}(1 + \cos 2\theta) + \frac{F_2}{2}(1 - \cos 2\theta)$$

Results are shown in Fig. 2-11 for the AVCO 5505 boron-epoxy material system.[22] Intrinsic strength properties for this material system are given below:

$X_1^T = 1.88 \times 10^5$ psi, $X_2^T = 9.0 \times 10^3$ psi $S_6 = 10.0 \times 10^3$ psi

$X_1^c = 3.61 \times 10^5$ psi, $X_2^c = 45.0 \times 10^3$ psi

Substituting these data into eq (2.117), one can calculate the constants of the quadratic failure criterion

$F_1 = 2.55 \times 10^{-6}\,(\text{psi})^{-1}$ $F_{11} = 1.47 \times 10^{-11}\,(\text{psi})^{-2}$

$F_2 = 8.89 \times 10^{-5}\,(\text{psi})^{-1}$ $F_{22} = 2.27 \times 10^{-9}\,(\text{psi})^{-2}$

$F_{66} = 1.00 \times 10^{-8}\,(\text{psi})^{-2}$

The results shown in Fig. 2-11 for boron-epoxy indicate that the off-axis tensile strength of the tensile coupons of orientation 5 deg–40 deg is governed by the inplane shear strength (S_6) while for the angles 40 deg–90 deg transverse tensile strength (X_2^T) controls. It should be noted that the quadratic interaction criterion shows close agreement with experimental data for all angles 0 deg–90 deg while the maximum stress criterion is unconservative in its prediction of strength in the region 5 deg–60 deg. This behavior is consistent with an expected strength reduction in this region resulting from the interaction of stress components in determining strength.

Hashin and Rotem[23] suggested an approach to lamina strength in which fiber dominated failure is separated from matrix dominated

---

**Note that* F_{12} *and* F_6 *are taken as zero.*

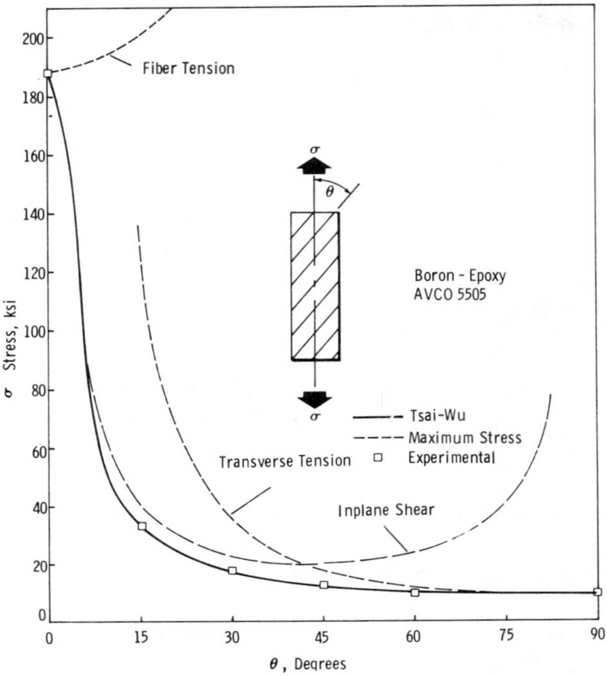

Fig. 2-11—Off-axis tensile strength (Pipes and Cole[22])

failure. In particular, the maximum stress criterion, eq (2.114), is applied to axial strength (X_1^T, X_1^c), and the following combined stress criterion is applied to the matrix dominated failures

$$\left(\frac{\sigma_2}{X_2^T}\right)^2 + \left(\frac{\sigma_6}{S_6}\right)^2 = 1$$
$$\left(\frac{\sigma_2}{X_2^c}\right)^2 + \left(\frac{\sigma_6}{S_6}\right)^2 = 1 \qquad (2.125)$$

In eq (2.125) it is assumed that $S_6^+ = S_6^-$. The quadratic interaction criterion, eq (2.116), can also be easily applied to matrix dominated failures with the result

$$\left(\frac{1}{X_2^T} - \frac{1}{X_2^c}\right)\sigma_2 + \frac{\sigma_2^2}{X_2^T X_2^c} + \left(\frac{\sigma_6}{S_6}\right)^2 = 1 \qquad (2.126)$$

Equation (2.126) has the advantage of incorporating differences in transverse tension and compression into one relationship. Again positive and negative shear strengths are assumed equal ($F_6 = 0$).

2.6.4 Laminate Strength Prediction

Prediction of laminate strength is accomplished by first determining the stress or strain distribution within each lamina of the laminate followed by a systematic application of a given strength criterion. Since the laminate stress state is a function of the laminate configuration, loading and laminae material properties, the elastic stress distribution can be determined only when all of these factors are known. Furthermore, inelastic behavior of the lamina invalidates the stress field predicted by the elastic analysis presented thus far. Therefore, it should not be surprising to find that the maximum strain criterion has received widespread adoption since the distribution of the strain through the laminate thickness is known when the surface strain components are known, regardless of material nonlinearity. The procedure discussed by Petit and Waddoups[24] is typical of this approach and is important because it provides for further loading beyond first ply failure. Many laminate strength criteria are based on first ply failure, which appears to provide a conservative estimate of laminate failure. Laminate failure criteria should also consider residual stress and strain.

2.7 ELASTIC STRESS CONCENTRATIONS

The elastic stress field in a homogeneous and orthotropic material is a function of the geometry of through-the-thickness material discontinuities such as circular holes or notches. The following interior discontinuity geometries will be examined in this section: the elliptical hole, the circular hole, and center crack.

2.7.1 Elliptical Hole

Consider a laminated plate of balanced, symmetric construction containing an elliptical hole with the major and minor axes parallel to the x and y material axes, respectively. The plate is of infinite extent with a remote uniaxial stress applied parallel to the y-axis, as shown in Fig. 2-12. In terms of laminated plate theory, the far field stress resultant is denoted by N_y^∞, yielding the average stress

$$\sigma_y^\infty = N_y^\infty/h = \text{constant} \tag{2.127}$$

In terms of strength reduction due to the presence of the elliptical hole, the effective normal stress along the major axis is of particular interest, that is, the normal stress

$$\sigma_y(x,0) = N_y(x,0)/h \tag{2.128}$$

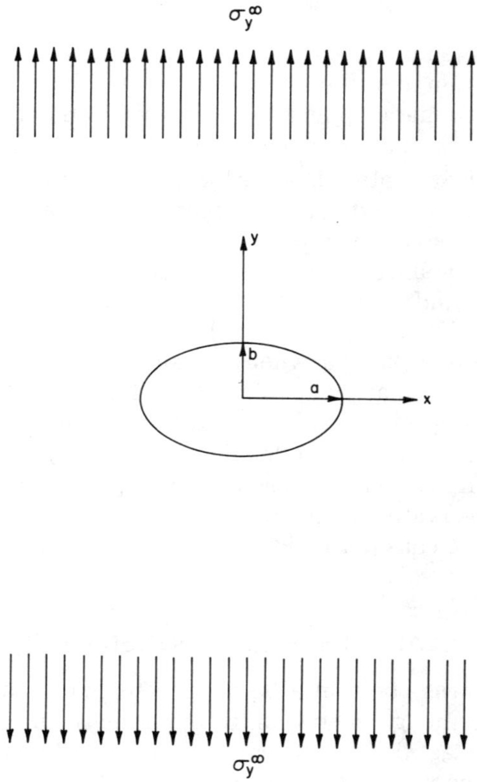

Fig. 2-12—Infinite plate containing an elliptical hole under remote uniform tension

The precise function corresponding to eq (2.128) can be determined using a complex variable technique.[25] Although the solution can be expressed in closed form, it is a rather complicated function of complex variables.

The stress concentration factor, K_T^∞, is also of interest and can be expressed in a simplified form.[21]

$$K_T^\infty = \frac{\sigma_y(a,0)}{\sigma_y^\infty} = \frac{N_y(a,0)}{N_y^\infty} = 1 + n\left(\frac{a}{b}\right) \tag{2.129}$$

where a and b are the length of the major and minor semi-axes of the ellipse, respectively, and

$$n = \sqrt{\frac{2}{A_{11}}\left(\sqrt{A_{11}A_{22}} - A_{12} + \frac{A_{11}A_{22} - A_{12}^2}{2A_{66}}\right)} \tag{2.130}$$

The equation for n may also be expressed in terms of the effective elastic moduli of the orthotropic laminate.[21]

$$n = \sqrt{2\left(\sqrt{\frac{E_y}{E_x}} - \nu_{yx} + \frac{E_y}{2G_{xy}}\right)} \tag{2.131}$$

For an isotropic material, $n = 2$ and eq (2.129) becomes

$$K_T^{\infty} = 1 + 2\left(\frac{a}{b}\right) \tag{2.132}$$

2.7.2 Circular Hole

The stress field in the vicinity of a circular hole of radius, R, in an orthotropic laminate under uniaxial loading, σ_y^{∞} can be determined from the solution to the elliptical hole as the special case where $a = b = R$. Again, the effective normal stress $\sigma_y(x,0)$ is a complicated expression of complex variables. This function can be approximated very accurately, however, from the relationship.[26]

$$\sigma_y(x,0) = \frac{\sigma_y^{\infty}}{2}\left\{2 + \left(\frac{R}{x}\right)^2 + 3\left(\frac{R}{x}\right)^2 - (K_T^{\infty} - 3)\left[5\left(\frac{R}{x}\right)^6 - 7\left(\frac{R}{x}\right)^8\right]\right\},$$

$$x \geq R \tag{2.133}$$

Using eq (2.129) for the case $a = b$ yields

$$K_T^{\infty} = 1 + n \tag{2.134}$$

For the isotropic case, $n = 2$ and eq (2.134) yield the classic value $K_T^{\infty} = 3$, yielding the following form of eq (2.133).

$$\sigma_y(x,0) = \frac{\sigma_y^{\infty}}{2}\left[2 + \left(\frac{R}{x}\right)^2 + 3\left(\frac{R}{x}\right)^4\right], \quad x \geq R \tag{2.135}$$

which is an exact solution for the isotropic case.

2.7.3 Center Notch

The stress in the vicinity of a center crack of length $2c$ in an orthotropic laminate under uniaxial loading, σ_y^{∞} can be derived as a special case of an elliptical hole with $a = c$ and $b \rightarrow 0$. It is easily seen from eq (2.129) that the stress concentration factor becomes unbounded, yielding a stress singularity at the crack tip. For this case, the exact expression for the normal stress ahead of the crack is given by[25]

$$\sigma_y(x,0) = \frac{\sigma_y^\infty x}{\sqrt{x^2 - c^2}}, \quad x > c \tag{2.136}$$

Note that eq (2.136) is independent of material properties. This is only true if the laminate is orthotropic (or, of course, isotropic).

Because of the stress singularity at the crack tip, the concept of a stress concentration factor is meaningless. For values of x close to the crack tip, however, eq (2.136) can be approximated by the relationship[27]

$$\sigma_y(x,0) = \frac{K^\infty}{\sqrt{2\pi(x-c)}} \tag{2.137}$$

where K^∞ is the stress intensity factor given by

$$K^\infty = \sigma_y^\infty \sqrt{\pi c} \tag{2.138}$$

Thus, for a sharp center crack, the concept of a stress concentration factor is replaced by a stress intensity factor which forms the basis of classical fracture mechanics. Uniaxial tension is referred to as mode I and the notation K_I^∞ is often used in conjunction with eq (2.138). The exact stress distribution, eq (2.136), in terms of the stress intensity factor is given by

$$\sigma_y(x,0) = \frac{K^\infty x}{\sqrt{\pi c(x^2 - c^2)}}, \quad x > c \tag{2.139}$$

2.8 STRENGTH OF NOTCHED COMPOSITE LAMINATES

Because of their importance in design applications, laminated, continuous-fiber reinforced polymeric matrix composites containing through-the-thickness cutouts in the form of sharp cracks and circular holes have been a subject of considerable study. A number of models have been proposed for notched composites in which some details of the damage region adjacent to the notch are included.[28-32] These models are very useful for gaining insight on the mechanisms associated with failure of notched composites. Two simplified approaches are of practical interest for predicting the notched strength of laminated composites. One approach utilizes concepts of linear elastic fracture mechanics (LEFM), while the second approach is based on the stress distribution adjacent to the notch. These approaches are generally applicable to laminates containing a number of 0 deg plies parallel to the load direction (filament dominated laminates).

2.8.1 Fracture Mechanics Criteria

It has been shown[33-35] that the strength of a laminated composite containing a sharp center crack of length $2c$ subjected to a uniaxial load can be predicted from LEFM. In particular,

$$K_Q = \sigma_N^\infty = \sqrt{\pi c} \tag{2.140}$$

where σ_N^∞ is the notched strength for a laminate of infinite width and K_Q is the critical stress intensity factor. The parameter K_Q is a function of c for small crack lengths and asymptotically approaches a constant value for large crack lengths. The concept of a plastic zone has been utilized in metals[36] to explain the increasing value of K_Q with increasing crack length. Although most polymeric matrix composites fail in a brittle manner, a damage zone does develop which is analogous to the plastic zone. Using this concept in conjunction with eq (2.140) yields

$$K_c = \sigma_N^\infty \sqrt{\pi(c + c_o)} \tag{2.141}$$

where K_c is the fracture toughness and c_o is an inherent flaw size. The term inherent flaw size is used because the unnotched strength, σ_o, of a composite laminate is given by eq (2.141) for the case of vanishing c, i.e.,

$$K_c = \sigma_o \sqrt{\pi c_o} \tag{2.142}$$

It should be noted that c_o does not physically refer to an inherent crack, but to a crack-like damage region which develops prior to ultimate failure. From a cursory examination of eqs (2.140) and (2.141) it can be seen that

$$K_c = \lim_{c \to \infty} K_Q \tag{2.143}$$

Combining eqs (2.141) and (2.142) yields the failure criterion

$$\sigma_N^\infty = \sigma_o \sqrt{1 - \xi_1} \tag{2.144}$$

where

$$\xi_1 = \frac{1}{c + c_o} \tag{2.145}$$

Experimental data[34-37] have shown that the tensile strength of laminated composites containing a circular hole depends on the hole size. It is obvious that such a phenomenon cannot be predicted by a classical stress concentration factor (SCF). Waddoups, Eisenmann, and Kaminski[37] used LEFM to explain the hole size effect. In particular, they assumed that a damage region which developed adjacent to the hole, perpendicular to the load direction, could be modeled as through-the-

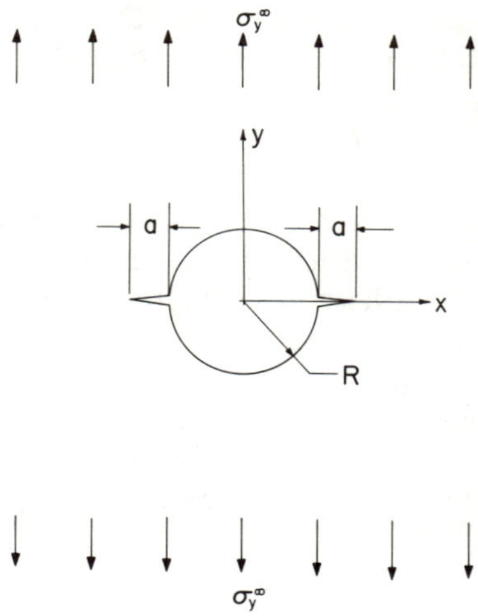

Fig. 2-13—Fracture mechanics model for circular hole

thickness cracks of length a, as illustrated in Fig. 2-13. A solution for this problem has been developed by Bowie[38] for an infinite isotropic plate with the result

$$K_c = \sigma_N^\infty \sqrt{\pi a}\, f(a/R) \qquad (2.146)$$

where R is the hole radius. The function $f(a/R)$ has been tabulated in Ref. 39. For a vanishing hole radius, $a/R \rightarrow \infty$, $f(a/R)$ approaches unity. Thus,

$$K_c = \sigma_o \sqrt{\pi a} \qquad (2.147)$$

Combining eqs (2.146) and (2.147) yields

$$\sigma_N^\infty = \sigma_o / f(a/R) \qquad (2.148)$$

It is assumed that a is independent of hole radius. Thus, eq (2.148) predicts the notch strength to be a function of absolute hole size. Numerical difficulties are encountered with this approach, however, because the function $f(a/R)$ has not been tabulated for orthotropic materials.

2.8.2 Stress Fracture Criteria

An alternate approach to LEFM for predicting uniaxial notch strength was proposed in Refs. 34 and 36, and is based on the stress

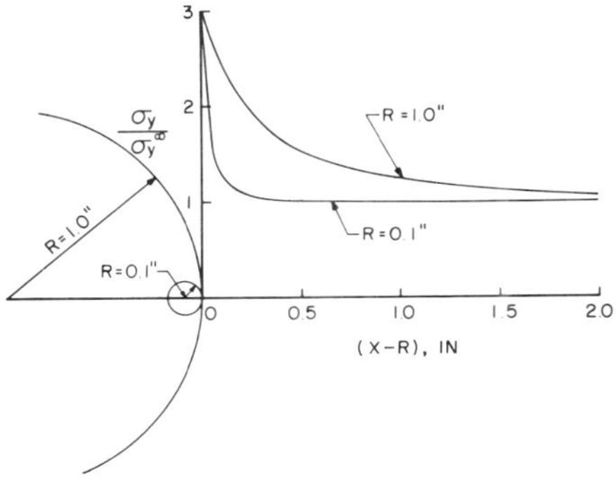

Fig. 2-14—Effect of hole radius on adjacent normal stress distribution in an isotropic material (Whitney and Nuismer[34])

distribution adjacent to the notch, perpendicular to the load. In this approach, the explanation of the hole size effect was based on the difference that exists in the normal stress distribution ahead of the hole for different sized holes, as shown for an isotropic material in Fig. 2-14. It is seen that, although all sized holes have the same stress concentration factor, the normal stres perturbation from a uniform stress state is considerably more concentrated near the hole boundary in the case of the smaller hole. Thus, intuitively, one might expect the plate containing the smaller hole to be the stronger of the two, since a greater opportunity exists in this case to redistribute high stresses. Through-the-thickness cracks were also considered in Refs. 34 and 35, where it was found that the crack size effect could be explained by considering the exact elasticity solution for the normal stress ahead of a crack rather than the singular term of the asymptotic expansion (see Fig. 2-15). Two

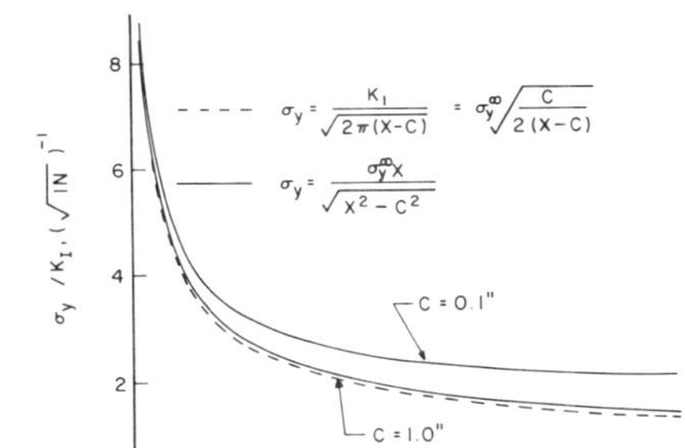

Fig. 2-15—Comparison of approximate and exact solution for stress distribution ahead of a center crack (Whitney and Nuismer[34])

stress criteria were proposed, the "point stress criterion" and the "average stress criterion."

In the "point stress criterion" failure is assumed to occur when the stress σ_y at some fixed distance, d_o, ahead of the notch becomes equal to the unnotched tensile strength, σ_o, of the material. For a circular hole, failure occurs when

$$\sigma_y(x,0)\Big|_{x = R + d_o} = \sigma_o \tag{2.149}$$

Using the approximate stress distribution for a circular hole in an infinite orthotropic plate, eq (2.133), in conjunction with eq (2.149) yields

$$\sigma_N^\infty = \frac{2\sigma_o}{[2 + \xi_2^2 + 3\xi_2^4 - (K_T^\infty - 3)(5\xi_2^6 - 7\xi_2^8)]} \tag{2.150}$$

where

$$\xi_2 = \frac{R}{R + d_o} \tag{2.151}$$

Note that for very large holes, ξ_2 approaches unity and the strength reduction is given by the SCF, that is,

$$\sigma_N^\infty = \frac{\sigma_o}{K_T^\infty} \tag{2.152}$$

Considering the center cracked geometry of Fig. 2-15, the exact anisotropic elasticity solution for the normal stress ahead of a crack of length $2c$ in an infinite anisotropic plate under uniform axial tension, σ_y^∞, is given by eq (2.136)

$$\sigma_y(x,0) = \frac{\sigma_y^\infty x}{\sqrt{x^2 - c^2}} = \frac{K_I^\infty x}{\sqrt{\pi c(x^2 - c^2)}}, \quad x>c \tag{2.153}$$

where K_I^∞ is the stress intensity factor, given by eq (2.136). Using eq (2.153) in conjunction with eq (2.149) with R replaced by c, one obtains

$$\sigma_N^\infty = \sigma_o\sqrt{1 - \xi_3^2} \tag{2.154}$$

where

$$\xi_3 = \frac{c}{c + d_o} \tag{2.155}$$

The second stress failure criterion proposed in Refs. 34 and 35, the "average stress criterion," assumes failure to occur when the average stress value of σ_y over some fixed distance, a_o, ahead of the notch first reaches the unnotched tensile strength of the material, that is, for the circular hole when

$$\frac{1}{a_o} \int_R^{R+a_o} \sigma_y(x,0)\,dx = \sigma_o \tag{2.156}$$

Using eq (2.156) in conjunction with eq (2.133) yields

$$\sigma_N^\infty = \frac{2\sigma_o(1-\xi_4)}{[2-\xi_4^2-\xi_4^4+(K_T^\infty-3)(\xi_4^6-\xi_4^8)]} \tag{2.157}$$

where

$$\xi_4 = \frac{R}{R+a_o} \tag{2.158}$$

Again for very large hole sizes, eq (2.157) reduces to the SCF.

Using eq (2.153) in conjunction with eq (2.156) with R replaced by c, one finds

$$\sigma_N^\infty = \sigma_o \sqrt{\frac{1-\xi_5}{1+\xi_5}} \tag{2.159}$$

where

$$\xi_5 = \frac{c}{c+a_o} \tag{2.160}$$

The predicted crack size effect on the measured value of the critical stress intensity factor, K_Q, can be observed by writing eqs (2.154) and (2.159) in the form

$$K_Q = \sigma_o \sqrt{\pi c(1-\xi_3^2)} \tag{2.161}$$

$$K_Q = \sigma_o \sqrt{\pi c \frac{(1-\xi_5)}{(1+\xi_5)}}. \tag{2.162}$$

respectively. These results are obtained by using the relationship between K_Q and σ_N^∞ given in eq (2.140). For large crack lengths, K_Q asymptotically approaches a constant value in both eqs (2.161) and (2.162), yielding a predicted fracture toughness given by

$$K_c = \sigma_o \sqrt{2\pi d_o} \tag{2.163}$$

$$K_c = \sigma_o \sqrt{\pi \frac{a_o}{2}} \tag{2.164}$$

Using the inherent flaw model as given in eq (2.142) in conjunction with eqs (2.163) and (2.164), one obtains a relationship between c_o and the stress fracture criteria parameters, d_o and a_o. In particular,

$$c_o = 2d_o \tag{2.165}$$

$$c_o = \frac{a_o}{2} \tag{2.166}$$

Using the inherent flaw model as given by eq (2.141) in conjunction with eqs (2.159) and (2.166), one obtains eq (2.164). Thus, the average stress criterion in conjunction with the inherent flaw model yields a fracture toughness which is independent of crack size. The same procedure in conjunction with the point stress criterion does not yield a relationship for K_c which is independent of crack length. Numerical results, however, show K_c to be relatively constant for all values of c.

It is assumed in the development of the stress fracture criteria that d_o and a_o are independent of notch size. Although results are presented only for a circular hole and center crack, the criteria can be extended to any notch geometry in which the stress distribution adjacent to the notch is known. These criteria are most effective when used to predict the failure of notched laminates which are fiber dominated (failure mode is characterized by fiber breakage rather than matrix fracture).

2.9 EDGE EFFECTS IN COMPOSITE LAMINATES

The presence of a boundary layer in the vicinity of the free edge of a composite laminate has been well established in recent studies.[49-55] It is important to recognize the presence of this boundary layer and its effect upon the surface strain distribution in that region, as well as the deleterious influence of the associated interlaminar stresses upon the strength of composite laminates.

2.9.1 Interlaminar Stress Boundary Layer

Thus far, the state of stress within each lamina of a multidirectional laminate has been assumed to be planar, wherein the interlaminar stress components vanish. This assumption is accurate for interior regions removed from laminate geometric discontinuities, such as free edges. However, there exists a boundary layer near the laminate free edge where the state of stress is three-dimensional. The boundary layer is the region wherein stress transfer between laminae is accomplished through the action of interlaminar stresses. The width of the boundary layer is a function of the elastic properties of the laminae, the laminae fiber orientations, and laminate geometry. A simple rule of thumb proposed in Ref. 40 for hard-polymer-matrix composites is that *the boundary layer is approximately equal to the laminate thickness.*

The primary consequences of the laminate boundary layer are delamination induced failures which initiate within this region and distortions of surface strain and deformations due to the presence of interlaminar stress components. Failures which initiate in the boundary-layer region of a finite-width test specimen may not yield data which accurately represent the true laminate strength.

Although previous experience has allowed the experimentalist to monitor strains at the edge of discontinuities in order to determine stress concentrations, the presence of the boundary layer precludes determination of subsurface behavior by surface strain measurements in this region.

2.9.2 First Mode Mechanism

The mechanism of interlaminar stress transfer for the $\pm\theta$ deg (angle-ply) laminate consists of laminae of only $+\theta$ deg and $-\theta$ deg fiber orientations. For a laminate subjected to axial load (x direction) only, the state of stress within each of the lamina at interior regions of the laminate is given below:

$$\begin{aligned} \sigma_x(\theta) &= \sigma_x(-\theta) = \overline{Q}_{11}\epsilon_x^\circ + \overline{Q}_{12}\epsilon_y^\circ \\ \tau_{xy}(\theta) &= -\tau_{xy}(-\theta) = \overline{Q}_{16}\epsilon_x^\circ + \overline{Q}_{26}\epsilon_y^\circ \\ \sigma_y(\theta) &= \sigma_y(-\theta) = 0 \end{aligned} \tag{2.167}$$

Hence, the inplane shear stresses within the $+\theta$ and $-\theta$ laminae are of equal magnitude but opposite sign. In addition, the shear stress, τ_{xy}, must vanish along free edges $y/b = 1.0$ (see Fig. 2-16). Hence, the interlaminar shear stress τ_{xz} is required to accomplish the sign change in the shear stress τ_{xy} at the interface and to equilibrate the inplane shear stress within the laminae.

Both approximate solutions of the exact equatons of elasticity and approximate analytic solutions for the finite-width angle-ply laminate have been developed.[40,50] A comparison between solution of displacement-equilibrium equations by finite differences and an approximate analytic solution is given in Refs. 48 and 50. Results for the inter-

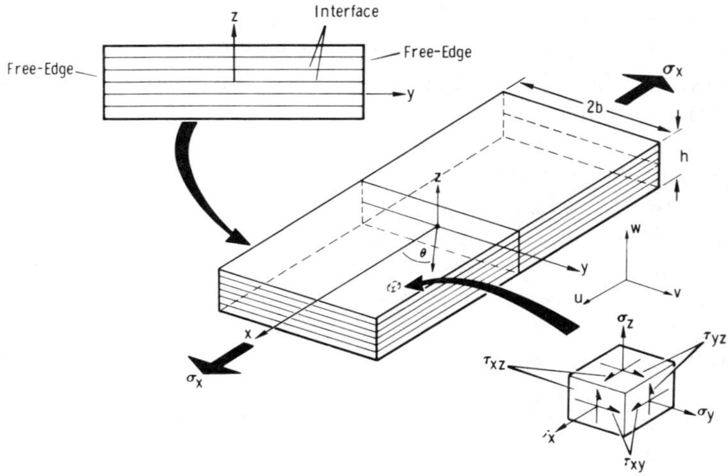

Fig. 2-16—Laminate geometry (Pagano and Pipes[42])

laminar shear stress τ_{xz} at the interface between laminae of +45 deg and −45 deg fiber orientation are presented in Fig. 2-17. These results show that the interlaminar shear stress is maximum at the free edge where it appears to grow without bound (Fig. 2-18). In addition, the displacement of the laminate surface in the direction of the load increases near the free edge, resulting in local inplane shear strains in this region as shown in Fig. 2-19. Finally the distribution of axial displacement and

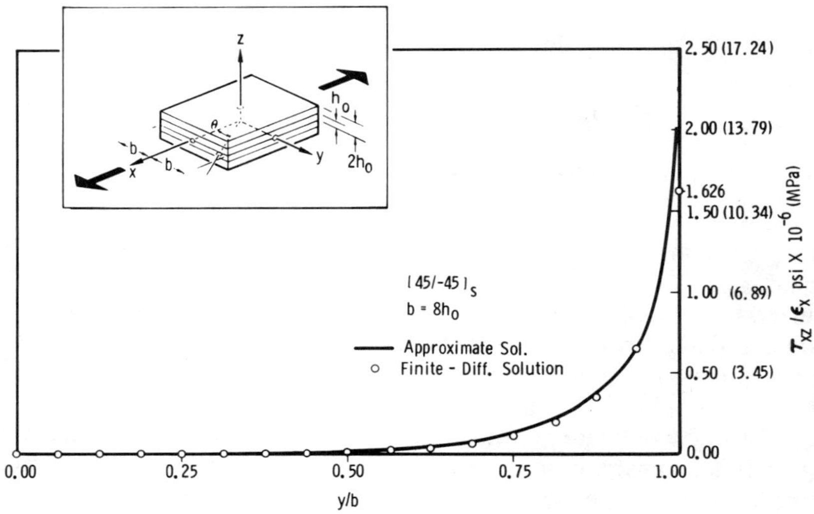

Fig. 2-17—Interlaminar shear stress distribution at z/h_o = 1.0 (Pipes and Pagano[50])

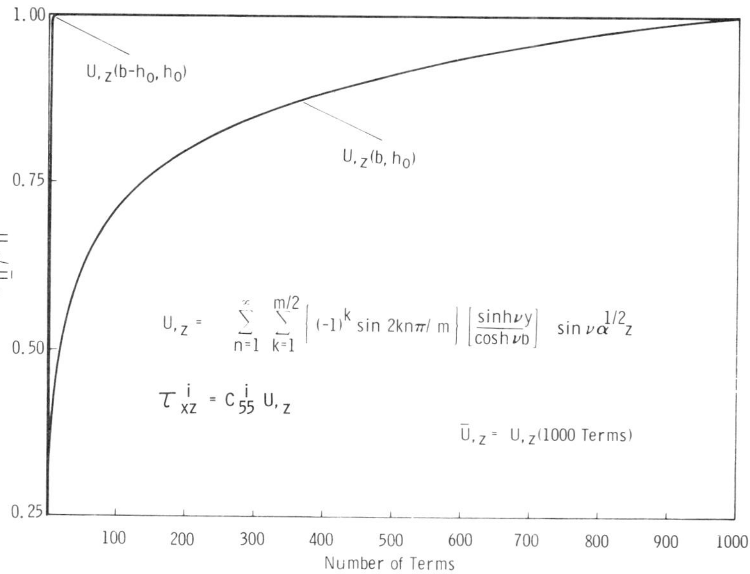

$$U_{,z} = \sum_{n=1}^{\infty} \sum_{k=1}^{m/2} \left\{ (-1)^k \sin 2kn\pi/m \right\} \left[\frac{\sinh \nu y}{\cosh \nu b} \right] \sin \nu \alpha^{1/2} z$$

$$\tau^i_{xz} = C^i_{55} U_{,z}$$

Fig. 2-18—Divergence of $U_{,z}$ at the free edge (Pipes and Pagano[50])

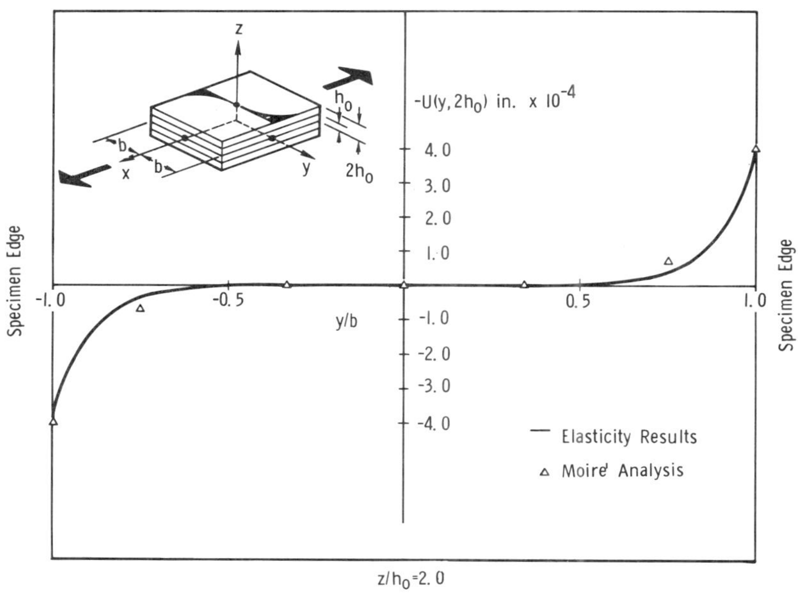

Fig. 2-19—Axial displacement distribution at the laminate surface (Pipes and Daniel[56])

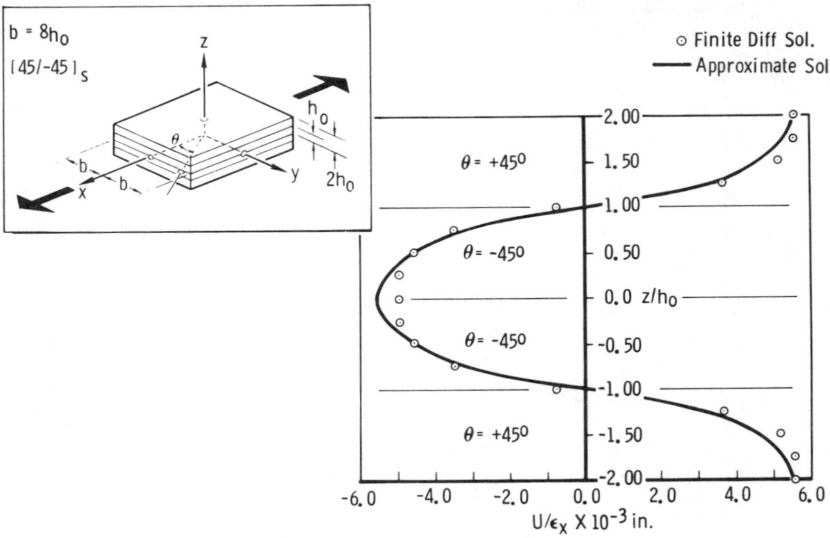

Fig. 2-20—Axial displacement at y/b = 1.0 (Pipes[48])

interlaminar shear stress through the laminate thickness is shown in Figs. 2-20 and 2-21. These results show that the maximum interlaminar shear stress occurs at the interface between the +45 deg and −45 deg layers.

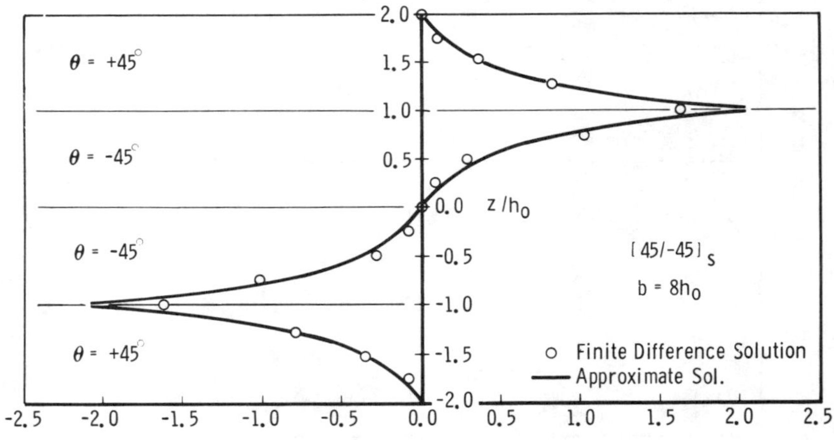

Fig. 2-21—Interlaminar shear stress distribution at y/b = 1.0 (Pipes[48])

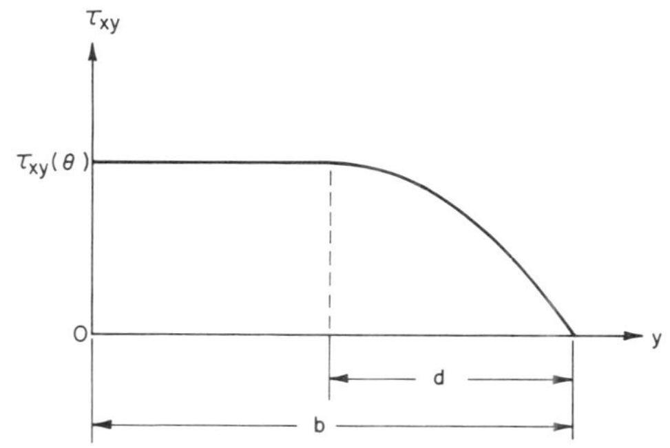

$$\tau_{xy} = \tau_{xy}(\theta), \qquad 0 \le y \le b-d$$

$$\tau_{xy} = \frac{\tau_{xy}(\theta)}{d^2}(y-b)(b-2d-y), \qquad b-d \le y \le b$$

Fig. 2-22—Assumed distribution of τ_{xy} across laminate width

Equilibrium equations of classical theory of elasticity for a laminate of thickness h require

$$\tau_{xz}(z) = -\int_{-h/2}^{z} \frac{\partial \tau_{xy}}{\partial y}\, dn \tag{2.168}$$

If a simplified distribution of the inplane shear stress, τ_{xy}, as shown in Fig. 2-22, is assumed across the laminate width, a relation can be derived for the maximum interlaminar shear stress, τ_m, in terms of the inplane shear stress, $\tau_{xy}(\theta)$, of the $+\theta$ layer as determined from laminated plate theory (see paragraph 2.6.1). For the $[\theta/-\theta]_{ns}$ class of laminates

$$\tau_m = \frac{h}{2nd}\, \tau_{xy}(\theta) \tag{2.169}$$

where d is the boundary layer width. Using results presented in paragraph 2.6.1, eq (2.169) becomes

$$\tau_m = \frac{h}{2nd}\left[\frac{A_{22}\overline{Q}_{16} - A_{12}\overline{Q}_{26}}{A_{11}A_{22} - A_{12}^2}\right]\bar{\sigma}_x \tag{2.170}$$

where $\bar{\sigma}_x$ is the average axial stress applied to the laminate or N_x/h. For a boundary layer of $d = h$,

$$\tau_m = \frac{1}{2n}\left[\frac{A_{22}\overline{Q}_{16} - A_{12}\overline{Q}_{26}}{A_{11}A_{22} - A_{12}^2}\right]\bar{\sigma}_x \tag{2.171}$$

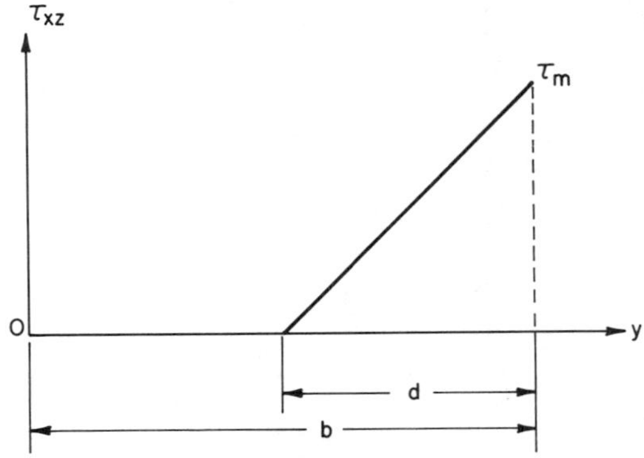

$$\tau_{xz} = 0\ , \qquad 0 \le y \le b - d$$

$$\tau_{xz} = \frac{\tau_m}{d}\,(y - b + d)\ , \qquad b - d \le y \le b$$

Fig. 2-23—Distribution of τ_{xz} across laminate width, resulting from assumed distribution of τ_{xy}

The distribution of the interlaminar shear stress, τ_{xz}, across the width of the laminate resulting from the assumed distribution of τ_{xy} is shown in Fig. 2-23.

The dependence of the interlaminar shear stress upon the fiber orientation, θ, of the laminae of an angle-ply laminate is characterized by the shear coupling terms, $\overline{Q}_{16}$ and $\overline{Q}_{26}$, in eqs (2.170) and (2.171). The variation of these terms with the angle, θ, for several material systems reveals the fiber orientations of maximum and minimum interlaminar shear stress.

2.9.3 Second Mode Mechanism

The mechanism of interlaminar stress transfer for the bidirectional laminate, $[0_i/90_j]_s$ will be termed the mode II mechanism. The bidirectional laminate* consists of laminae of either 0 deg or 90 deg fiber orientation. For a bidirectional laminate subjected to axial load (x-direction) only, the state of stress for each lamina at interior regions of the laminate is given as:

**This kind of laminate is often known as a cross-ply laminate.*

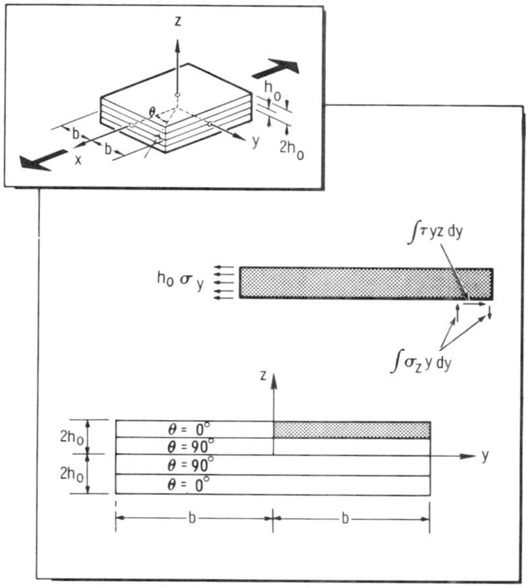

Fig. 2-24—Bidirectional laminate interlaminar stress transfer mechanism (Pipes[48])

$$\begin{aligned}
\sigma_x(0) &= Q_{11}\epsilon_x^o + Q_{12}\epsilon_y^o \\
\sigma_x(90) &= Q_{22}\epsilon_x^o + Q_{12}\epsilon_y^o \\
\sigma_y(0) &= -\sigma_y(90) = Q_{12}\epsilon_x^o + Q_{22}\epsilon_y^o \\
\tau_{xy}(0) &= \tau_{xy}(90) = 0
\end{aligned} \tag{2.172}$$

In contrast to the state of stress for the angle-ply laminate, the inplane shear stress component vanishes for the bidirectional laminate due to the fact that the shear coupling stiffness terms $\overline{Q}_{16}$ and $\overline{Q}_{26}$ vanish at orientations of 0 deg or 90 deg. However, a mismatch between Poisson's ratios of the 0 deg and 90 deg laminae of the bidirectional laminate lead to equal but opposite-sign transverse stresses. Thus the interlaminar stress, τ_{yz}, is required at the interface in order to accomplish the transverse stress (σ_y) sign change between the laminae. Further, the force vectors acting on the surface laminae due to the σ_y and τ_{yz} stresses are not colinear (see Fig. 2-24) and hence result in a couple whose magnitude is given by

$$\text{Couple} = \sigma_y h_0^2/2 \tag{2.173}$$

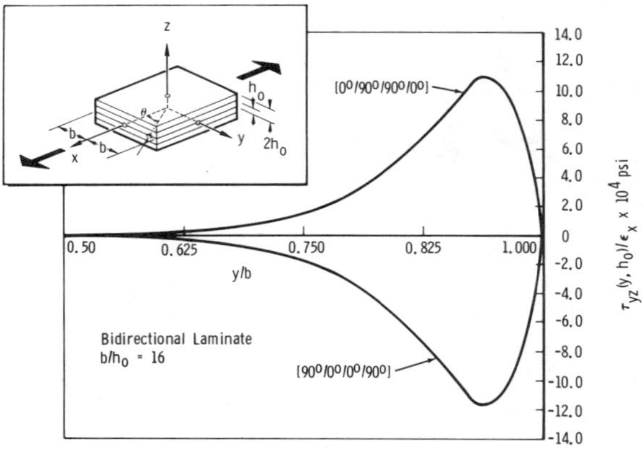

Fig. 2-25—Distribution of the interlaminar shear stress at the interface (Pipes[48])

The couple is reacted by the interlaminar normal stress component, σ_z. The distribution of the interlaminar normal stress must therefore result in zero vertical force vector while producing a couple equal in magnitude to that given in eq (2.173).

Solutions of the equations of elasticity for the bidirectional laminate have been obtained by numerical methods.[47] In addition, finite-element techniques have also been employed to establish the distribution of interlaminar stresses at the interface of a bidirectional laminate.[44] These studies have revealed distributions of the interlaminar shear and normal stresses as shown in Figs. 2-25 and 2-26.

If simplified distributions of the interlaminar stresses are assumed which satisfy equilibrium, it is possible to derive a simple relationship between the interlaminar normal stress and the applied axial stress, $\bar{\sigma}_x$. Consider the approximate distribution of σ_z given in Fig. 2-27. If the coordinate ξ is measured from the edge of the boundary layer toward the free edge, then the distribution is a constant $\sigma_m/5$ over the region $0<\xi/d<0.667$, changes sign at $\xi/d = 0.723$ and reaches a maximum of σ_m at $\xi/d = 1.000$. This distribution satisfies the pure couple requirement of equilibrium. The maximum interlaminar normal stress, σ_m, can now be related to applied axial stress, $\bar{\sigma}_x$. Following the treatment of Pagano,[49] σ_m can be related to the transverse stress, σ_y. For a laminate of thickness h

$$\sigma_m(z) = \frac{90}{7d^2} \int_z^{h/2} \sigma_y(\eta)(\eta - z)\,d\eta \qquad (2.174)$$

But the transverse stress, σ_y, can be written in terms of the applied axial stress, $\bar{\sigma}_x$, and the laminae stiffness terms

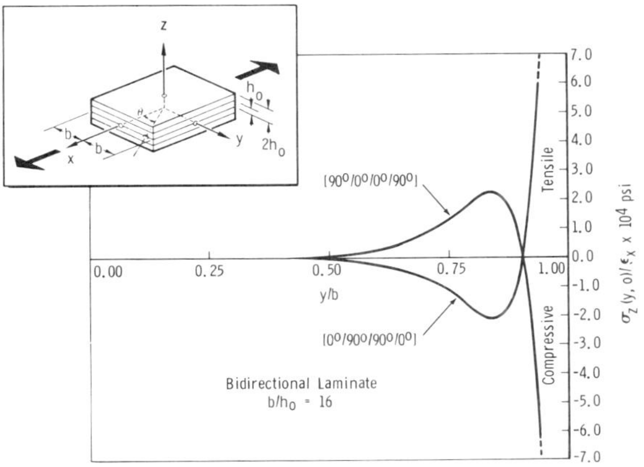

Fig. 2-26—Distribution of interlaminar normal stress at the midplane (Pipes[48])

$$\sigma_y(z) = \left[\frac{A_{22}\overline{Q}_{12}(z) - A_{12}\overline{Q}_{22}(z)}{A_{11}A_{22} - A_{12}^2}\right] \bar{\sigma}_x h \tag{2.175}$$

Thus eqs (2.174) and (2.175) can be combined to yield

$$\sigma_m(z) = \frac{90}{7Ad}\,\bar{\sigma}_x \int_z^{h/2} [A_{22}\overline{Q}_{12}(\eta) - A_{12}\overline{Q}_{22}(\eta)](\eta - z)d\eta \tag{2.176}$$

where $A = A_{11}A_{22} - A_{12}^2$.

Equation (2.176) allows determination of the maximum interlaminar normal stress at any position, z, throughout the thickness of the laminate. Hence, it is possible to examine alternative laminate stacking sequences in order to minimize the interlaminar normal stress.

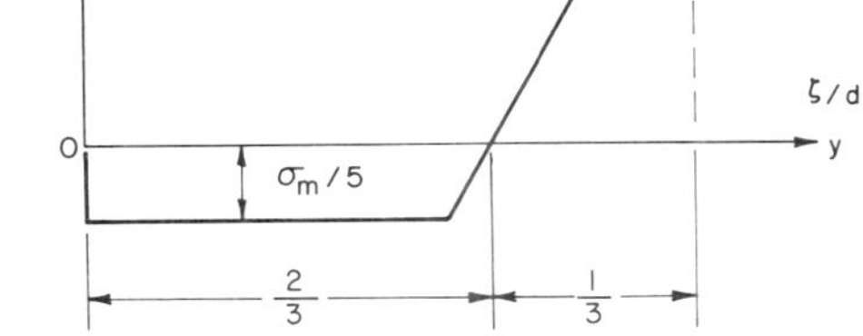

Fig. 2-27—Assumed distribution of σ_z across laminate width (Pagano and Pipes[49])

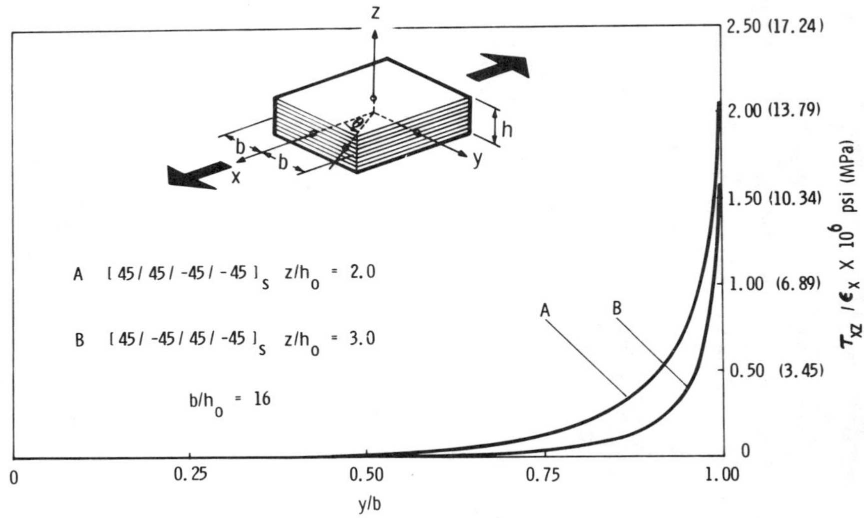

Fig. 2-28—Influence of stacking sequence upon the interlaminar shear stress (Pipes and Pagano[50])

2.9.4 *Influence of Stacking Sequence on Interlaminar Stresses*

Laminate stacking sequence may have a significant influence upon the interlaminar stresses within the boundary layer. For the mode I mechanism, stacking sequences in which layers of common orientation are lumped together, thus effectively producing larger layer thicknesses, h_o, result in an increase in τ_{xz}. This result is confirmed in eqs (2.171), as well as in results presented in Fig. 2-28 from Ref. 50.

Perhaps the most important of the interlaminar stresses in initiating interlaminar failure is the interlaminar normal stress, σ_z. This is the case primarily because most fiber composites possess interlaminar strength properties in tension which are inferior to their interlaminar compressive strengths. Hence, stacking sequences which produce tensile interlaminar normal stresses at the edge should be expected to be more delamination prone than stacking sequences which yield compressive interlaminar stresses at the edge. Experimental data which support this conclusion may be found in Refs. 46, 49 and 55.

REFERENCES

1. "Laminate Orientation Code," Section 4.0.5, *Advanced Composites Design Guide,* Vol. 4, Materials, 3rd Edition, Air Force Flight Dynamics Laboratory (Dec. 1975).

2. Hashin, Z., "Analysis of Properties of Fiber Composites with Anisotropic Constituents," *Journal of Applied Mechanics,* Vol. 46, Vol. 3, 543-550 (Sept. 1979).
3. Halpin, J.C., "Stiffness and Expansion Estimates for Oriented Short Fiber Composites," *Journal of Composite Materials,* Vol. 3, 732-734 (1969).
4. Ashton, J.E., Halpin, J.C. and Petit, P.H., *Primer on Composite Materials: Analysis,* Technomic Publishing Co., Inc., Stamford, CT (1969).
5. Whitney, J.M., "Elastic Moduli of Unidirectional Composites with Anisotropic Filaments," *Journal of Composite Materials,* Vol. 1, 188-193 (1967).
6. Schapery, R.A., "Thermal Expansion Coefficients of Composite Materials Based on Energy Principles," *Journal of Composite Materials,* Vol. 2, 380-404 (1968).
7. Pister, K.S. and Dong, S.B., "Elastic Bending of Layered Plates," *Journal of the Engineering Mechanics Division,* American Society of Civil Engineers, Vol. 85, 1-10 (Oct. 1959).
8. Reissner, E. and Stavsky, Y., "Bending and Stretching of Certain Types of Heterogeneous Aeolotropic Elastic Plates," *Journal of Applied Mechanics,* Vol. 28, 402-408 (Sept. 1961).
9. Ashton, J.E. and Whitney, J.M., *Theory of Laminated Plates,* Technomic Publishing Co., Inc., Stamford, CT (1970).
10. Hahn, H.T. and Pagano, N.J., "Curing Stresses in Composite Laminates," *Journal of Composite Materials,* Vol. 9, 91-106 (1975).
11. Hahn, H.T., "Residual Stresses in Polymer Matrix Composite Laminates," *Journal of Composite Materials,* Vol. 10, 266-278 (1976).
12. Hahn, H.T. and Kim, R.Y., "Swelling of Composite Laminates," *Advanced Composite Materials—Environmental Effects,* ASTM STP 658, J.R. Vinson, Editor, American Society for Testing and Materials, 98-120 (1978).
13. Hill, R., "A Theory of the Yielding and Plastic Flow of Anisotropic Metals," *Proceedings of the Royal Society,* London, Series A, Vol. 193, 281-297 (1948).
14. Azzi, V.D. and Tsai, S.W., "Anisotropic Strength of Composites," *Experimental Mechanics,* Vol. 5, No. 9, 283-288 (Sept. 1965).
15. Tsai, S.W. and Azzi, V.D., "Strength of Laminated Composite Materials," *AIAA Journal,* Vol. 4, No. 2, 296-301 (Feb. 1966).
16. Hoffman, O., "The Brittle Strength of Orthotropic Materials," *Journal of Composite Materials,* Vol. 1, No. 2, 200-297 (April 1967).
17. Tsai, S.W. and Wu, E.M., "A General Theory of Strength for Anisotropic Materials," *Journal of Composite Materials,* Vol. 5, No. 1, 58-80 (Jan. 1970).
18. Goldenblat, I. and Kopnov, V.A., "Strength of Glass-Reinforced Plastics in The Complex Stress State," *Mekhanika Polimerov,* Vol. 1, 70-78 (1965); English translation: *Polymer Mechanics,* Vol. 1, 54-60 (1966), published by Faraday Press.
19. Narayanaswami, R. and Adelman, H.M., "Evaluation of the Tensor Polynomial and Hoffman Strength Theories for Composite Materials," *Journal of Composite Materials,* Vol. 11, No. 4, 366-377 (Oct. 1977).
20. Tsai, S.W. and Hahn, H.T., *Introduction to Composite Materials,* Technomic Publishing Company, Westport, CT (1980).
21. Tsai, S.W., "Strength Theories of Filamentary Structures," *Fundamental*

Aspects of Fiber Reinforced Plastic Composites, Edited by R.T. Schwartz and H.S. Schwartz, Wiley Interscience, New York, 3-11 (1968).

22. Pipes, R.B. and Cole, B.W., "On the Off-Axis Strength Test of Anisotropic Materials," *Journal of Composite Materials,* Vol. 7, No. 2, 246-256 (April 1973).
23. Hashin, Z. and Rotem, A., "A Fatigue Failure Criterion for Fiber Reinforced Materials," *Journal of Composite Materials,* Vol. 7, 448-464 (1973).
24. Petit, P.H. and Waddoups, M.E., "A Method of Predicting the Nonlinear Behavior of Laminated Composites," *Journal of Composite Materials,* Vol. 3, 2-19 (Jan. 1969).
25. Lekhnitskii, S.G., *Anisotropic Plates,* Translated from the Second Russian Edition by S.W. Tsai and T. Cheron, Gordon and Breach Science Publishers, New York (1968).
26. Konish, H.J. and Whitney, J.M., "Approximate Stresses in an Orthotropic Plate Containing a Circular Hole," *Journal of Composite Materials,* Vol. 9, No. 2, 157-166 (April 1975).
27. Sih, G.C., Paris, P.C. and Irwin, G.R., "On Cracks in Rectilinearly Anisotropic Bodies," *International Journal of Fracture Mechanics,* Vol. 1, 189-203 (1965).
28. Kulkarni, S.V., Rosen, B.W. and Zweben, C., "Load Concentration Factors for Circular Holes in Composite Laminates," *Journal of Composite Materials,* Vol. 7, 387-393 (July 1973).
29. Zweben, C., "Fracture Mechanics and Composite Materials: A Critical Analysis," *Analysis of the Test Methods for High Modulus Fibers and Composites,* ASTM STP 521, American Society for Testing and Materials, 65-97 (1973).
30. Wang, S.S., Mandell, J.F. and McGarry, F.J., "Three-Dimensional Solution for a Through-Thickness Crack with Crack Tip Damage in a Cross-Plied Laminate," *Fracture Mechanics of Composites,* ASTM STP 593, American Society for Testing and Materials, 61-85 (1975).
31. Nuismer, R.J. and Brown, Gary E., "Progressive Failure of Notched Composite Laminates Using Finite Elements," *Advances in Engineering Science,* 13th Annual Meeting Society of Engineering Science, NASA CP-2001, Vol. 1, 183-192 (1976).
32. Chun, Shun-Chin, Orringer, O. and Rainey, J.H., "Post-Failure Behavior of Laminates: II-Stress Concentration," *Journal of Composite Materials,* Vol. 11, 71-78 (Jan. 1977).
33. Konish, H.J., Jr. and Cruse, T.A., "Determination of Fracture Strength in Orthotropic Graphite/Epoxy Laminates," *Composite Reliability,* ASTM STP 580, American Society for Testing and Materials, Philadelphia, 490-503 (1975).
34. Whitney, J.M. and Nuismer, R.J., "Stress Fracture Criteria for Laminated Composites Containing Stress Concentrations," *Journal of Composite Materials,* Vol. 8, 253-275 (1974).
35. Nuismer, R.J. and Whitney, J.M., "Uniaxial Failure of Composite Laminates Containing Stress Concentrations," *Fracture Mechanics of Composites,* ASTM STP 593, American Society for Testing and Materials, Philadelphia, 117-142 (1975).

36. McClintock, F.A. and Irwin, G.R., "Plasticity Aspects of Fracture Mechanics," *Fracture Toughness Testing and Its Applications,* ASTM STP 381, American Society for Testing and Materials, Philadelphia, 84-113 (1964).
37. Waddoups, M.E., Eisenmann, J.R. and Kaminski, B.E., "Macroscopic Fracture Mechanics of Advanced Composite Materials," *Journal of Composite Materials,* Vol. 5, 446-454 (1971).
38. Bowie, O.L., "An Analysis of an Infinite Plate Containing Radial Cracks Originating from the Boundary of an Internal Circular Hole," *Journal of Mathematics and Physics,* Vol. 35, 60-71 (1956).
39. Paris, P.C. and Sih, G.C., "Stress Analysis of Cracks," *Fracture Toughness Testing and Its Applications,* ASTM STP 381, American Society for Testing and Materials, Philadelphia, 30-81 (1964).
40. Pipes, R.B. and Pagano, N.J., "Interlaminar Stresses in Composite Laminates Under Uniform Axial Extension," *Journal of Composite Materials,* Vol. 4, No. 4, 538-548 (Oct. 1970).
41. Puppo, A.H. and Evensen, H.A., "Interlaminar Shear in Laminated Composites Under Generalized Plane Stress," *Journal of Composite Materials,* Vol. 4, No. 2, 204-221 (April 1970).
42. Pagano, N.J. and Pipes, R.B., "Influence of Stacking Sequence on Laminate Strength," *Journal of Composite Materials,* Vol. 5, No. 1, 50-57 (Jan. 1971).
43. Isakson, G. and Levy, A., "Finite Element Analysis of Interlaminar Shear in Fibrous Composites," *Journal of Composite Materials,* Vol. 5, No. 2, 273-276 (April 1971).
44. Rybicki, E.F., "Approximate Three-Dimensional Solutions for Symmetric Laminates Under Inplane Loading," *Journal of Composite Materials,* Vol. 5, No. 3, 354-360 (July 1971).
45. Whitney, J.M., "Free-Edge Effects in the Characterization of Composite Materials," *Analysis of the Test Methods for High Modulus Fibers and Their Composites,* ASTM STP 521, American Society for Testing and Materials, Philadelphia, 167-180 (1973).
46. Whitney, J.M. and Browning, C.E., "Free-Edge Delamination of Tensile Coupons," *Journal of Composite Materials,* Vol. 6, No. 2, 300-303 (April 1972).
47. Oplinger, D.W., "Edge Effects in Angle-Ply Composite," *AMMRC TR 71-62,* Army Materials and Mechanics Research Center, Watertown, MA (1971).
48. Pipes, R.B., "Solution of Certain Problems in the Theory of Elasticity for Laminated Anisotropic Systems," Ph.D. Dissertation, University of Texas at Arlington (March 1972).
49. Pagano, N.J. and Pipes, R.B., "Some Observations on the Interlaminar Strength of Composite Laminates," *International Journal of Mechanical Sciences,* Vol. 15, 679-688 (1973).
50. Pipes, R.B. and Pagano, N.J., "Interlaminar Stresses in Composite Laminates—An Approximate Elasticity Solution," *Journal of Applied Mechanics,* Vol. 41, No. 3, 668-672 (Sept. 1974).
51. Pagano, N.J., "On the Calculation of Interlaminar Normal Stress in Composite Laminate," *Journal of Composite Materials,* Vol. 8, No. 1, 65-82 (Jan. 1974).
52. Hsu, P.W. and Herakovich, C.T., "A Perturbation Solution for Interlaminar

Stresses in Bidirectional Laminates," *Composite Materials: Testing and Design* Fourth Conference), ASTM STP 617, American Society for Testing and Materials, Philadelphia, 296-316 (1977).

53. Wang, A.S.D. and Crossman, F.W., "Some New Results on Edge Effects in Symmetric Composite Laminates," *Journal of Composite Materials,* Vol. 11, No. 1, 92-106 (Jan. 1977).
54. Wang, A.S.D. and Crossman, F.W., "Thermally Induced Edge Stresses in Symmetric Composite Laminates," *Journal of Composite Materials,* Vol. 11, No. 3, 300-312 (July 1977).
55. Whitney, J.M. and Kim, R.Y., "Effect of Stacking Sequence on the Notched Strength of Laminated Composites," *Composite Materials: Testing and Design* (Fourth Conference), STP 617, American Society for Testing and Materials, Philadelphia, 229-242 (1977).
56. Pipes, R.B. and Daniel, I.M., "Moiré Analysis of the Interlaminar Shear Edge Effect in Laminated Composites," *Journal of Composite Materials,* Vol. 5, No. 2, 255-259 (April 1971).

3

EXPERIMENTAL STRAIN ANALYSIS

3.1 ELECTRICAL RESISTANCE STRAIN GAGES

3.1.1 Introduction

Electrical resistance strain gages are very sensitive for measuring deformations in composite materials. Strain measurement is based on the electrical resistance change of the gage bonded to the part undergoing deformation. The strain is given by

$$\epsilon = \frac{1}{S_g}\left(\frac{\Delta R}{R}\right) \tag{3.1}$$

where S_g is the gage factor, a calibration constant which is a function of the gage alloy and the backing materials, R the gage resistance, and ΔR the change in resistance.[1] Commercially available foil gages come in a variety of sizes, configurations, and combinations (rosettes) and are very suitable for composite materials. Gages with gage lengths as small as 0.38 mm (0.015 in.) have been applied to the curved edges of cutouts in eight-ply laminates and near crack tips.[2-11] Techniques have been developed for embedding foil gages between plies during the lamination process and residual strains during curing of the laminate have been measured.[12-18] Strain gages have also been used to measure wave propagation in composites and to obtain strain data at very high rates of loading.[19-22]

3.1.2 Temperature Compensation

In many strain gage applications, test conditions are not isothermal. Hence, techniques must be employed to compensate for free thermal strains that the material exhibits due to a change in temperature. Although this phenomenon is also encountered in isotropic materials, the anisotropy of composite materials requires that temperature compensation of these materials receive special considerations. When the ambient temperature changes by ΔT four important effects result:[1]

(1) The strain sensitivity of the metal alloy of the gage changes with temperature.
(2) The substrate (composite) elongates by $\alpha_\theta \Delta T$.
(3) The gage elongates by $\beta \Delta T$.
(4) The resistance of the gage changes by $\gamma \Delta T$ due to the influence of the temperature coefficient of resistivity of the gage material.

The combined effect of all these factors produces the following temperature-induced resistance change:

$$\left(\frac{\Delta R}{R}\right)_{\Delta T} = (\alpha_\theta - \beta) S_g \Delta T + \gamma \Delta T \tag{3.2}$$

where α_θ is the coefficient of thermal expansion of the composite in the direction parallel to the gage axis, β is the thermal coefficient of expansion of the gage and γ is the temperature coefficient of resistivity of the gage material. For experiments where the temperature variation exceeds $\pm 10°$ C, it is necessary to take into account the temperature dependence of the gage factor S_g, i.e., the change in strain sensitivity of the gage alloy.

The coefficient α_θ may be expressed as a function of the principal coefficients of thermal expansion of the laminate, α_x and α_y as shown previously in eq (2.93).

$$\begin{aligned} \alpha_\theta &= m^2 \alpha_x + n^2 \alpha_y \\ m &= \cos\theta \\ n &= \sin\theta \end{aligned} \tag{3.3}$$

It should be noted that, like α_θ, the temperature-induced change in resistance of a gage depends strongly on its orientation with respect to the principal material axes of the laminate.

If the coefficients of thermal expansion of the strain gage and the composite material are equal ($\alpha_\theta = \beta$), the first term of eq (3.2) vanishes and only a nonzero coefficient of resistivity, γ, can produce a change in gage resistance due to the change in temperature. This approach has been adopted by strain gage manufacturers to produce temperature-compensated gages by perfectly matching the coefficient of thermal expansion of known base materials (steel and aluminum) while holding the temperature coefficient of resistivity at zero. However, since thermal expansion of a composite material is a function of layup and orientation, θ, it is not possible to develop temperature-compensated gages for composite materials, in general.

Two methods are recommended for temperature compensation of strain gages on composite materials. The first approach employs the features of the Wheatstone bridge which utilizes an unloaded dummy gage to cancel the change in resistance of the actual gage due to a change in temperature. This is accomplished in the same manner as for isotropic materials provided that the orientation of the dummy gage with respect to the principal material axes is identical to that of the active gage. Any difference in alignment will result in a difference in thermal expansion coefficients of the actual material tested and the dummy substrate material. Of course, it is obvious that the dummy substrate material must possess identical thermal expansion properties to those of the actual material.

The second method for temperature compensation uses an identical gage bonded on a reference material of known thermal expansion. One reference gage is sufficient for any number of similar gages applied to a composite and exposed to the same temperature changes. Then, the true thermal strain in the composite is given by

$$\epsilon_t = \epsilon_a - \epsilon_r + \epsilon_{tr} \tag{3.4}$$

where ϵ_t is the true strain, ϵ_a the apparent (uncorrected) strain in the composite, ϵ_r the apparent strain in the reference material, and ϵ_{tr} the known ($\alpha_r \Delta T$) thermal expansion of the reference material. Reference materials used normally are aluminum oxide, fused quartz, and titanium silicate with coefficients of thermal expansion of $6.8 \times 10^{-6}K^{-1}$ (3.8 μin./in./° F), $0.7 \times 10^{-6}K^{-1}$ (0.4 μin./in./° F) and $0.03 \times 10^{-6}K^{-1}$ (0.017 μin./in./° F), respectively. Figure 3-1 shows the thermal response of strain gages temperature-compensated for steel mounted on aluminum oxide and fused quartz. To minimize the purely thermal response of the gage it is preferable to use gages with zero-expansion temperature compensation ($\beta \simeq 0$). The thermal response of such a gage mounted on a titanium silicate specimen is illustrated in Fig. 3-2.

3.1.3 Transverse Gage Sensitivity

The gage factor given by the manufacturer is defined as:[1]

$$S_g = \frac{\Delta R/R}{\epsilon_x} \quad \text{when } \epsilon_y = -0.285\,\epsilon_x \tag{3.5}$$

The procedure for obtaining S_g consists of mounting a gage on a steel coupon of Poisson's ratio $\nu = 0.285$ and subjecting the specimen to a uniaxial stress of σ_x. It is apparent that if the gage factor above is used to determine strain in any strain field other than the one under which

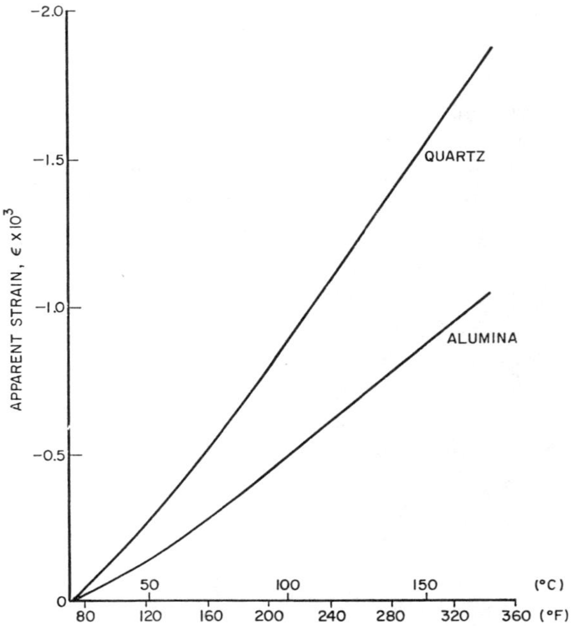

Fig. 3-1—Outputs of strain gages attached to quartz and aluminum oxide as a function of temperature [coefficients of thermal expansion for quartz and alumina are 0.7×10^{-6} K^{-1} (0.4×10^{-6} in./in./° F) and 6.8×10^{-6} K^{-1} (3.8×10^{-6} in./in./° F), respectively]

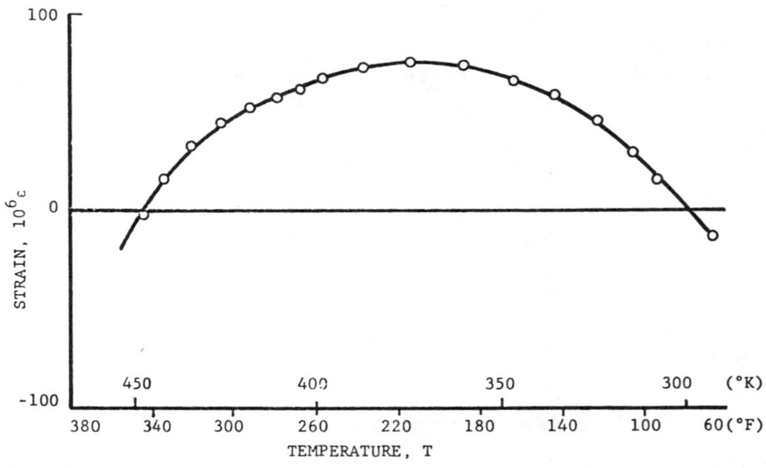

Fig. 3-2—Apparent strain as a function of temperature of WK-00-125TM-350 gage bonded on titanium silicate

S_g was determined, an error will result. The magnitude of this error can vary from insignificant to appreciable, depending on the case.

It is possible to correct the error by taking the cross-sensitivity factor of the gage into consideration. This factor is defined as

$$K = \frac{S_A}{S_T} \tag{3.6}$$

where S_A and S_T are the axial and transverse strain sensitivities defined as

$$S_A = \frac{\Delta R/R}{\epsilon_y} \qquad \text{when } \epsilon_x = 0 \tag{3.7}$$

and

$$S_T = \frac{\Delta R/R}{\epsilon_x} \qquad \text{when } \epsilon_y = 0 \tag{3.8}$$

For commonly used gages the cross-sensitivity factor K ranges between ± 0.01 and ± 0.02.

Taking the cross-sensitivity factor into consideration, the true strains at a point along two mutually perpendicular directions are obtained from the apparent strains as follows:

$$\begin{aligned} \epsilon_x &= \frac{1 - \nu_o K}{1 - K^2} (\epsilon_x' - K\epsilon_y') \\ \epsilon_y &= \frac{1 - \nu_o K}{1 - K^2} (\epsilon_y' - K\epsilon_x') \end{aligned} \tag{3.9}$$

where ϵ_x, ϵ_y are the actual strains, ϵ_x' and ϵ_y' are the apparent strains, and $\nu_o = 0.285$ is Poisson's ratio of the gage calibration material. The transverse sensitivity correction becomes appreciable if the strain to be measured is small compared to the strain in the perpendicular direction. This situation is encountered frequently in composites.

In the case of Poisson's ratio determination the true value is related to the uncorrected value as follows:

$$\begin{aligned} \nu &= -\frac{\epsilon_x}{\epsilon_y} = -\frac{\epsilon_x' - K\epsilon_y'}{\epsilon_y' - K\epsilon_x'} \\ &= \frac{\nu' + K}{1 + \nu' K} \simeq \nu' + (1 - \nu'^2)K \\ &\simeq \nu' \pm 0.02 \end{aligned} \tag{3.10}$$

where

ν = actual Poisson's ratio
ν' = uncorrected Poisson's ratio

The correction above is meaningful for the determination of the minor Poisson's ratio ν_{21} in unidirectional composites.

3.1.4 Strain Gage Circuits

In static applications each gage element is connected to one arm of a Wheatstone bridge circuit with three equal resistors. The voltage output of such a gage is

$$\Delta E = V \epsilon S_g \frac{r}{(1+r)^2} \tag{3.11}$$

where

V = excitation voltage
ϵ = strain
r = R_2/R_1, ratio of resistance of two adjacent arms of the Wheatstone bridge.

In applications involving a number of gages, a number of Wheatstone bridges are connected in parallel and excited by one common power supply. Within each bridge potentiometers are provided for span voltage (individual gage excitation voltage) and balance control. These signal conditioning circuits are normally used with digital data acquisition systems incorporating a digital voltmeter, a scanner, and printer.

In special cases in which the lead wires to the gages are long or in which they are partially exposed to high temperature variations, techniques such as the three-wire compensation are recommended.[1]

In dynamic applications standard potentiometric circuits are used. The transient strain signals are recorded on oscilloscopes or similar instruments.

3.1.5 Embedded Strain Gage Techniques

Composite laminates are amenable to embedment of strain gages between various plies during the fabrication process. This allows for strain measurements in the interior of composite laminates, monitoring of strains during curing and at elevated temperatures. Techniques have been developed and applied to the measurement of subsurface strains in boron/epoxy, boron/polyimide, graphite/epoxy, graphite/polyimide and glass/epoxy laminates.[12-18] It was demonstrated that conventional foil gages when embedded do not produce local thickening of the specimen, do not reduce the strength of the material, and do not affect the average mechanical properties. Valid strain readings were obtained up to failure of the material.[12]

The techniques in question consist primarily of embedding foil strain gages with attached leads between the plies of the composite during lamination, and recording the output during curing and subsequent thermal or mechanical loading. The primary requirements are that the integrity and stiffness of the laminate remain unaltered by the embedment, and that the strain gage be electrically insulated from the conducting fibers.

In the case of glass/epoxy composites there are no insulation problems. Conventional open-face foil gages with ribbon leads attached are simply laid on the ply with the desired orientation. The matrix resin serves to bond the gage. No additional cement is necessary.

In the case of boron/epoxy composites slightly different techniques are used. An area of the scrim cloth equal to the area of the gage is removed from the ply where the gage is to be applied. This prevents local thickening since both the scrim cloth and the gage are approximately 25 μm (0.001 in.) thick. The gages used are fully encapsulated in polyimide with the attached ribbon leads coated to fully insulate them from the conducting boron fibers (Micro-Measurements QA Series). In addition, since the ribbon lead coating is not always available or effective, the leads are sandwiched between strips of 13 μm (0.0005 in.) thick polyimide (Kapton sheet). Figure 3-3 shows a cross section through a $[0/\pm 45/\bar{0}]_s$ boron/epoxy specimen with embedded gages.* Figure 3-4 shows strain gages, with ribbon leads attached, located on an interior ply of the same laminate.

The same procedure above, except for the absence of scrim cloth, is used with graphite/epoxy composites. Extra precautions are taken to insulate the gages and leads from the highly conductive graphite fibers.

Special gages of a nickel-chromium alloy fully encapsulated in glass-reinforced epoxy-phenolic resin are used for the high-temperature curing boron/polyimide and graphite/polyimide composites (Micro-Measurements WK Series). The attached ribbon leads are sandwiched between thin sheets of polyimide (13 μm; 0.0005 in.) prior to embedment. The gage leads are silver-soldered to special nickel-clad copper wire with fiber-glass braid insulation. In order to prevent air leaks through the insulation and maintain vacuum during curing, it is necessary

*The overbar symbol denotes that the 0-deg ply is at the very center of the 7-ply laminate.

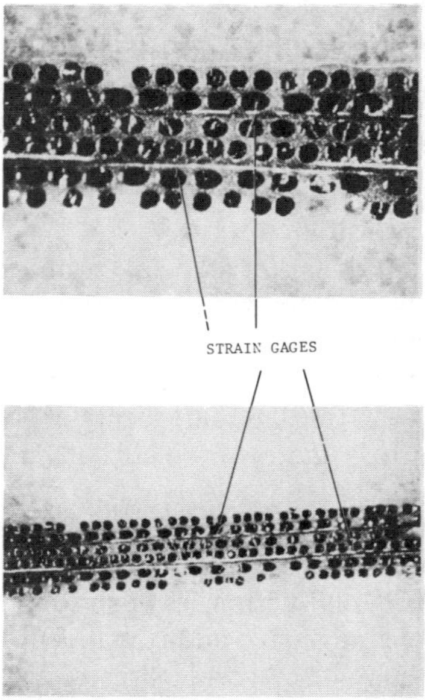

Fig. 3-3—Section through a gage location of an instrumented $[0/\pm 45/0]_s$ epoxy boron specimen showing the embedded three gage rosettes on the second and fourth ply. Note the twelve separate sections of a gage grid exposed below the second ply

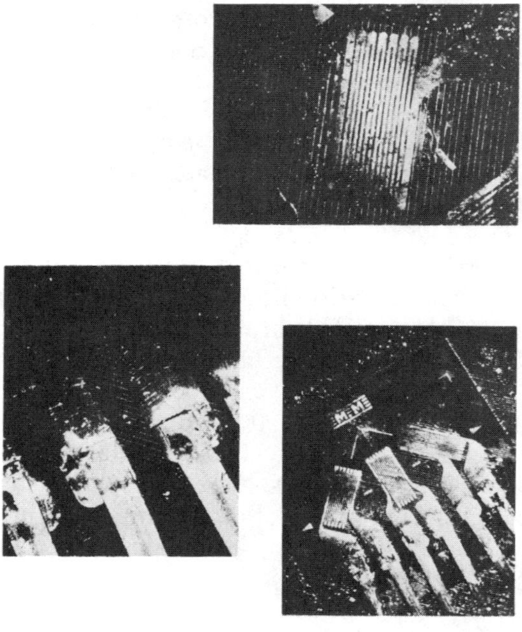

Fig. 3-4—Strain gages with ribbon lead wire located on epoxy boron composite

to bare the wires locally and embed them in the silicon rubber gasket used in bagging the specimens.

In the cases previously mentioned, all gages are completely wired before specimen curing and connected to a data acquisition system for monitoring during curing. The instrumented specimens are bagged with the gages wired and subjected to the prescribed curing and postcuring cycles in the autoclave.

Standard three-wire compensation techniques are used to compensate for appreciable resistance changes in the portions of the lead wires exposed to elevated temperatures inside the autoclave.

3.1.6 Measurement of Thermal Expansion

Strain gages have been used successfully for measuring thermal expansion in unidirectional and angle-ply laminates of various materials.[13-15,23,24] Strain gages record local deformations averaged over their gage length; therefore, they tend to reflect local material irregularities, inhomogeneities and flaws in a realistic composite. For this reason some variations may be seen between gage readings from different gages at different locations. These strain variations can be of the order $\pm 100\ \mu\epsilon$. Readings from surface gages may be different from those of embedded gages because of laminate bending. Realistic results are obtained by averaging readings from a number of embedded gages.

In one recent application of strain gages to measurement of thermal deformations,[15] encapsulated gages (WK-00-125TM-350, Option B-157) were embedded between the plies during laminate assembly. The attached ribbon leads were sandwiched between thin (0.013 mm; 0.0005 in.) polyimide strips. A thermocouple was also embedded in each laminate. The instrumented specimens, including a reference titanium silicate specimen, were subjected to the curing and postcuring cycles in the autoclave. Strain gage and thermocouple readings were taken throughout. Subsequently, the same specimens were subjected to a thermal cycle from room temperature to 444° K (340° F) and down to room temperature. Strain gages and thermocouples were recorded at 5.5° K (10° F) intervals. The true thermal strains were obtained from recorded apparent strains as discussed before.

Thermal strains measured on eight-ply unidirectional graphite/epoxy, Kevlar 49/epoxy, and *S*-glass/epoxy specimens are shown in Fig. 3-5. Both Kevlar 49/epoxy and graphite/epoxy exhibit negative thermal strains in the longitudinal (fiber) direction. The Kevlar 49/epoxy exhibits the largest positive transverse and negative longitudinal strains. The *S*-glass/epoxy undergoes the lowest thermal deformation in the transverse direction and the highest (positive) in the longitudinal direction.

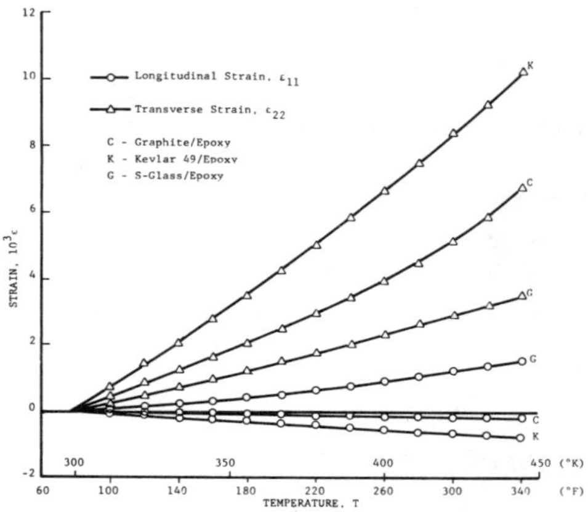

Fig. 3-5—Strains in $[0_8]$ unidirectional specimens as a function of temperature for three basic materials

Coefficients of thermal expansion computed from such data are tabulated in Table 3-1 for eight composite materials.[14,15,25] Coefficients are listed at room temperature, 297° K (75° F), and at the elevated temperature of 450° K (350° F). All graphite fiber composites exhibit negative thermal expansion in the fiber direction. The polyimide matrix composites do not show any variation of thermal coefficients with temperature. This is true at least up to the postcuring temperature of 589° K (600° F).

3.1.7 Determination of Residual Stresses

Lamination residual stresses in angle-ply laminates are produced during curing as a result of the anisotropic thermal deformations of the variously oriented plies. These stresses have been investigated recently both analytically and experimentally.[13-18,26-28] They are a function of many parameters, such as type of fiber and matrix, fiber volume ratio, ply orientation, curing temperature and other variables. They can reach values comparable to the transverse strength of the ply and thus induce cracking of that ply within the laminate. They are equilibrated with interlaminar shear stresses transmitted from adjacent plies and can cause delamination.

Residual stresses during curing have been measured in a variety of angle-ply laminates using embedded strain gage techniques.[13-15] Unidirectional and angle-ply specimens were instrumented with surface and embedded gages and thermocouples and the output recorded during curing and postcuring. The unidirectional specimen was used as a reference to determine the unrestrained stress-free thermal expansion of an individual ply.

Table 3-1 Thermal Expansion Coefficients of Unidirectional Composite Materials

Material	Longitudinal Coefficient of Thermal Expansion, α_{11}, $10^{-6}K^{-1}$ ($\mu\epsilon/°F$)				Transverse Coefficient of Thermal Expansion, α_{22}, $10^{-6}K^{-1}$ ($\mu\epsilon/°F$)			
	297° K	(75° F)	450° K	(350° F)	297° K	(75° F)	450° K	(350° F)
Boron/Epoxy (Boron/AVCO 5505)	6.1	(3.4)	6.1	(3.4)	30.3	(16.9)	37.8	(21.0)
Boron/Polyimide (Boron/WRD 9371)	4.9	(2.7)	4.9	(2.7)	28.4	(15.8)	28.4	(15.8)
Graphite/Epoxy (Modmor I/ERLA 4289)	−1.1	(−0.6)	3.2	(1.3)	31.5	(17.5)	27.0	(15.0)
Graphite/Epoxy (Modmor I/ERLA 4617)	−1.3	(−0.7)	−1.3	(−0.7)	33.9	(18.8)	83.7	(46.5)
Graphite/Polyimide (Modmore I/WRD 9371)	−0.4	(−0.2)	−0.4	(−0.2)	25.3	(14.1)	25.3	(14.1)
S-Glass/Epoxy (Scotchply 1009-26-5901)	3.8	(2.1)	3.8	(2.1)	16.7	(9.3)	54.9	(30.5)
S-Glass/Epoxy (S-Glass/ERLA 4617)	6.6	(3.7)	14.1	(7.9)	19.7	(10.9)	26.5	(14.7)
Kevlar/Epoxy (Kevlar 49/ERLA 4617)	−4.0	(−2.2)	−5.7	(−3.2)	57.6	(32.0)	82.8	(46.0)

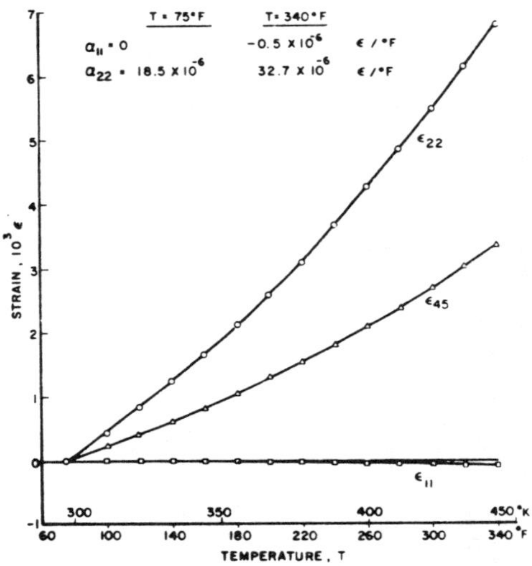

Fig. 3-6—Thermal strains in $[0_8]$ unidirectional graphite/epoxy specimen

It was found that apparent strains recorded during the heating stage of the curing cycle are not significant as they correspond to the fluid state of the matrix resin. Residual stress buildup occurs only upon solidification of the matrix near the peak curing temperature and during subsequent cooldown. Strains measured during the cooldown stage of curing as well as those measured during postcuring correspond to the thermal expansion of the laminate.

Thermal strains measured in a unidirectional graphite/epoxy laminate are shown in Fig. 3-6 with room temperature as the reference level. Thermal strains measured in a $[0_2/\pm 45]_s$ graphite/epoxy angle-ply laminate during the cooling stage of postcuring are shown in Fig. 3-7. The residual stresses induced in each ply correspond to the so-called restraint strains, i.e., the difference between the unrestrained thermal expansion of that ply (obtained from the unidirectional specimen) and the restrained expansion of the ply within the laminate (obtained from the angle-ply specimen). Restraint or residual strains in the 0- and 45-deg plies of the $[0_2/\pm 45]_s$ graphite/epoxy laminate are plotted in Figs. 3-8 and 3-9 as a function of temperature with room temperature taken as the zero strain reference level. The stress-free level can be shifted to 444° K (340° F), the peak curing temperature. Other investigators have claimed that the stress-free temperature level might be somewhat lower than the peak curing temperature as indicated by comparing experimental and theoretical results.[28] In the case in question the maximum residual strain at room temperature is 6.43×10^{-3} in the ± 45 deg

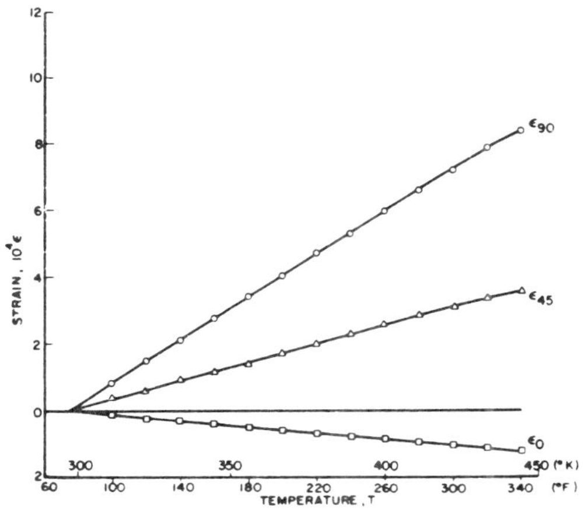

Fig. 3-7—Thermal strains in $[0_2/\pm 45]_s$ graphite/epoxy specimen

plies in the direction transverse to the fibers. The corresponding maximum residual strain in the 0-deg plies is 5.95×10^{-3}. These values are based on the assumption of a strain-free stress-free condition at 444° K (340° F).

Residual stresses in any given ply at any given temperature can be computed from the residual strains using the appropriate orthotropic constitutive relations. Assuming linear elastic behavior, these relations take the form

$$\sigma_{11} = \frac{E_{11}}{1 - \nu_{12}\nu_{21}} [\epsilon_{11} + \nu_{21}\epsilon_{22}]$$

$$\sigma_{22} = \frac{E_{22}}{1 - \nu_{12}\nu_{21}} [\epsilon_{22} + \nu_{12}\epsilon_{11}] \tag{3.12}$$

$$\sigma_{12} = 2G_{12}\epsilon_{12} = G_{12}\gamma_{12}$$

where subscripts 1 and 2 refer to the fiber direction and the inplane direction transverse to the fibers. The material properties in the constitutive relations above can vary with time and temperature. For linear elastic behavior the properties are time-independent and the temperature dependence is accounted for by using the properties at the temperature at which stresses are computed. For viscoelastic response (time-dependence) the whole time history enters into the stress determination.

Residual stresses at room temperature computed for the 0-deg and ±45-deg plies of the $[0_2/\pm 45]_s$ graphite/epoxy laminate at room temperature are tabulated in Table 3-2. The stress transverse to the fibers in the ±45-deg plies seems to exceed somewhat the measured trans-

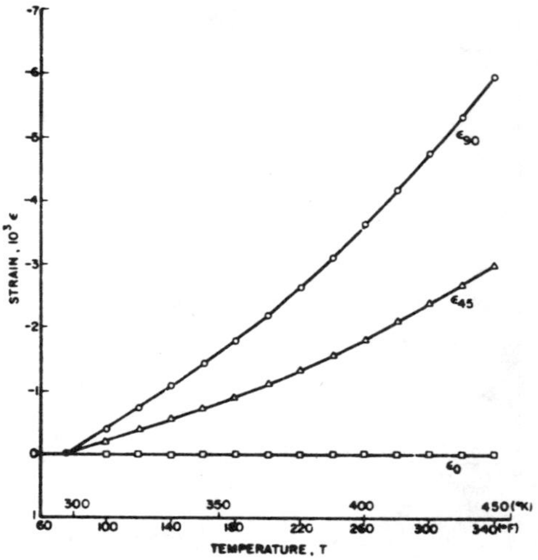

Fig. 3-8—Residual strains in 0-deg plies of $[0_2/\pm45]_s$ graphite/epoxy specimen

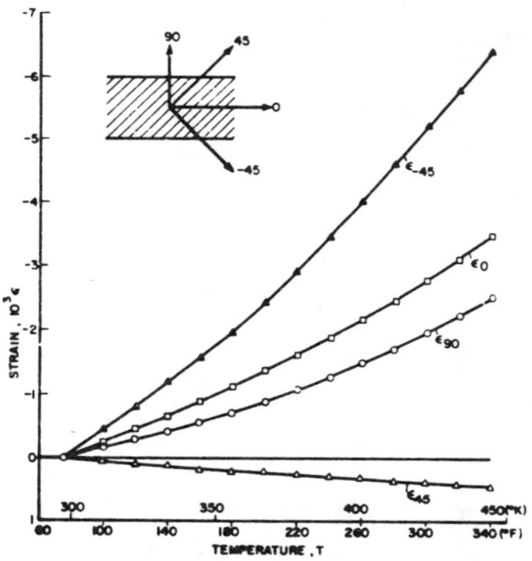

Fig. 3-9—Residual strains in 45-deg plies of $[0_2/\pm45]_s$ graphite/epoxy specimen

verse tensile strength of the unidirectional material which is 42 MPa (6.1 ksi). This means that these plies are probably damaged in their transverse direction upon completion of curing.

Comparable results for glass/epoxy and boron/epoxy show that residual tensile stresses exhaust a significant portion of the transverse tensile strength of the ply. In the case of Kevlar 49/epoxy computed residual stresses, assuming linear elastic behavior, far exceed the trans-

Table 3-2 Residual Stresses at Room Temperature in $[0_2/\pm 45]_s$ Graphite/Epoxy Laminate

Ply (deg)	Stress, MPa (ksi)		
	σ_{11}	σ_{22}	σ_{12}
0	23 (3.3)	42 (6.1)	0
±45	−52 (−7.5)	45 (6.5)	6 (0.9)

verse strength of the ply. It has been shown also that the amount of relaxation of residual stresses is fairly small.[29]

Further experimental work has been reported on the effects of laminate construction, ply stacking sequence and interply hybridization on residual stresses.[17,18]

3.2 MOIRÉ METHODS

3.2.1 *Introduction*

The moiré effect is an optical phenomenon observed when two closely spaced arrays of lines are superimposed and viewed with either transmitted or reflected light. If the two arrays consist of opaque parallel lines which are not identical in either spacing (pitch) or orientation, then interference between the two arrays occurs and moiré fringes are produced. The method yields full-field information of the inplane surface displacements from which strain and stress fields can be derived. Being strictly geometric in nature, the technique is independent of the anisotropy and inelasticity of the material.

The sensitivity of the moiré method depends on the density of the rulings used and on the magnitude of the deformation. Glass/epoxy composites in general, and graphite/epoxy and boron/epoxy laminates with a large number of ±45-deg plies, undergo sufficient deformation to yield analyzable moiré fringes using model arrays of 40 lines/mm (1000 ℓpi) and superimposing a similar master array during the test. However, this line density represents a practical upper limit for model arrays and, in many cases, the deformations to be measured are small. Therefore, the sensitivity of the moiré method must be enhanced by other means, such as fringe interpolation and multiplication techniques.

Interpolation between integral fringe orders has been achieved by correlating light intensity and fringe order,[30] by digital filtering techniques,[31] and by grid-shift techniques.[32,33] Moiré fringe multiplication techniques have been applied successfully to analysis of nonhomogeneous states of strain.[34-38]

The application of moiré techniques in general consists of three basic tasks: (1) application of rulings or grids to the specimen, (2) photographic recording of information in the form of fringe patterns or deformed grids, and (3) analysis of moiré data.

3.3.2 Conventional Methods

In conventional practice one array or a grid of lines is applied to the specimen surface by photographic, etching, transfer, or bonding techniques. A practical upper limit of line density is 40 lines/mm (1000 ℓpi). In some applications to translucent glass/epoxy laminates, film replicas of these arrays can be bonded directly to the specimen with the emulsion side outward, without any special preparation of the specimen surface. In the case of opaque materials, such as most composites, the surface is prepared with a smooth and partially reflective substrate. This has been done by two different techniques. In one case a thin layer of epoxy approximately 0.013 mm (0.0005 in.) thick is deposited and cured on the surface of the specimen and then a thin layer of aluminum (microns thick) is vacuum-deposited. In the other technique the specimen surface is cleaned and lightly sanded, coated with a layer of an epoxy primer and later with a coat of white polyurethane enamel.

After preparation of the surface by one of the two techniques mentioned, the line array or grid is either bonded or photoprinted onto this surface by the photoresist process (Kodak Photoresist, KPR).

During the test, a transparent reference array (master grid) of the same line density is placed near or in contact with the specimen array. Contact is maintained with a film of oil. As the specimen is loaded, its array of lines deforms to follow the surface displacements and a moiré interference pattern is formed. This pattern which is a measure of the displacement field, is recorded photographically. When the specimen array consists of a grid of dots, the reference or master array can be a similar grid of dots or an array of parallel lines. In the first case, two intersecting fringe patterns, corresponding to two families of mutually perpendicular displacements, are obtained. In the second case the two families of fringes are obtained and photographed successively by rotating the master array by 90 degrees.

A reference array does not have to be in contact with the specimen array everywhere. Other methods exist for obtaining moiré fringe patterns. Double exposure of the specimen array before and after loading and partial mirrors for optically combining the two images in a single photograph have been used. Sometimes the reference array is positioned in the camera back with provisions for rotating it by 90 degrees. One of the advantages of this arrangement is the flexibility in adjusting the relative pitch of the two arrays and obtaining finer or coarser moiré fringe patterns to increase the sensitivity of the method. A variation of the conventional moiré method, in which the moiré rulings are formed by the interference of two coherent (laser) beams, has also been used.[39]

The techniques mentioned previously are used for measuring in-plane displacements from which strains are computed. Moiré can also be used for measuring out-of-plane displacements. One technique used for this measurement is the shadow moiré. In this case a master array is placed in front of the specimen and illuminated by collimated light at an angle to the array. The moiré pattern formed by the interference of the master array and its shadow on the specimen is analyzed to determine the out-of-plane displacements.[40]

3.2.3 *Fringe-Multiplication Methods*

In many applications, especially with boron/epoxy and graphite/epoxy composites, strains are small even near fracture and it becomes necessary to enhance the sensitivity of the moiré method by means of fringe-multiplication techniques.[38] These techniques pose more severe requirements on procedures for applying and recording specimen grids, but they also allow the use of coarser specimen rulings.

In optimizing grid-deposition techniques, it was found that smoothness of the specimen surface is very important. A surface of glass-like smoothness is produced by casting a preferably nonshrinking plastic between the composite surface and a smooth plate. This is accomplished by sanding the specimen surface and applying carefully mixed and deaerated clear epoxy (0151 clear Hysol cement) and curing it under approximately 100 kPa (15 psi) pressure between the specimen and a polymethyl methacrylate plate. After a 24-hr room-temperature cure, the plate is removed leaving a very thin layer (less than 0.025 mm thick) of epoxy on the specimen with glass-like smoothness. If the original specimen surface is too rough, the operation above is carried in two steps. Subsequent to the surface preparation, the specimen is coated with a vacuum-deposited aluminum layer (microns thick) and a 20-dot/mm (500-dot/in.) grid is photoprinted on it by the KPR process. The specimen in this case exhibits nearly specular reflection instead of diffuse reflection, which makes it necessary to illuminate the specimen normally to its surface.

Application of fringe multiplication techniques requires the use of a special camera capable of recording the deformed specimen grid faithfully

Fig. 3-10—Camera for replication of specimen grids applied to a composite laminate

at a 1:1 magnification. Such a camera is shown in Fig. 3-10. The distance of the lens from the object is adjusted by means of three micrometers with anvils resting on a rigid plane parallel to that of the specimen. The exact location of the object plane is measured with three dial gages having their probes resting on the specimen. In operation, the correct location is adjusted within 0.0025 mm (0.0001 in.).

The illumination system consists of a 75-W incandescent lamp, heavily coated for diffused-light emission; a green filter; a 45-deg pellicle-type beam splitter to reflect light onto the specimen and then transmit a portion of the light from the specimen into the camera; and a Fresnel field lens between the lamp and beam splitter to direct light into the camera by forming an image of the lamp in the aperture of the camera lens. The specimen grid is faithfully recorded on high-resolution plates at no load and at various load levels.

The deformed grid replica is analyzed subsequently on an optical bench by illuminating it in series with a doubly-blazed reference (analyzing) grating at oblique incidence with a collimated monochromatic light beam. The line frequency of the analyzing grating is an integral multiple

β of the undeformed specimen grid frequency. The replica of the specimen grid and the analyzing grating act as diffraction gratings splitting the incident beam into many diffracted ones. The multiplication effect is achieved by collecting and isolating the beams of $-\frac{\beta}{2}$ and $+\frac{\beta}{2}$ diffraction order by means of a decollimating lens and an aperture plate. The resulting moiré fringes correspond to those obtained in a conventional experiment in which both arrays have the same pitch as the denser analyzing grating, i.e., a fringe multiplication of β is achieved. The underlying theory and the detailed process of fringe multiplication have been described by Post.[34-37]

3.2.4 Analysis of Moiré Fringes

Moiré patterns are related to the displacement field since they represent loci of points having the same component of displacement normal to the line array. In a fringe pattern produced by arrays of 40 lines/mm (1000 lines/in.), for example, each fringe represents a locus of points having a displacement of 1/40 mm (0.001 in.) relative to the points of a neighboring fringe. Line arrays parallel to the y- and x-axes of a specimen yield moiré fringe patterns corresponding to the families of displacement in the x- and y-directions, respectively (u and v families). Strains are obtained from the displacements by using the appropriate strain-displacement relations.

When strains are finite, the strain-displacement equations take the following form if deformations are referred to the original coordinates (Lagrangian definition)

$$\begin{aligned}
\epsilon_x &= \left[1 + 2\frac{\partial u}{\partial x} + \left(\frac{\partial u}{\partial x}\right)^2 + \left(\frac{\partial v}{\partial x}\right)^2\right]^{1/2} - 1 \\
\epsilon_y &= \left[1 + 2\frac{\partial v}{\partial y} + \left(\frac{\partial v}{\partial y}\right)^2 + \left(\frac{\partial u}{\partial y}\right)^2\right]^{1/2} - 1 \qquad (3.13) \\
\epsilon_{xy} &= \frac{1}{2}\arcsin\frac{\dfrac{\partial u}{\partial y} + \dfrac{\partial v}{\partial x} + \dfrac{\partial u}{\partial x}\dfrac{\partial u}{\partial y} + \dfrac{\partial v}{\partial x}\dfrac{\partial v}{\partial y}}{(1+\epsilon_x)(1+\epsilon_y)}
\end{aligned}$$

When strains are small, as is the case in most composite laminates, the products and powers of displacement derivatives in the expressions above can be neglected compared to the derivatives themselves. Then, the strain-displacement relations are reduced to the familiar form used in classical linear elasticity.

$$\epsilon_x = \frac{\partial u}{\partial x}$$

$$\epsilon_y = \frac{\partial v}{\partial y} \tag{3.14}$$

$$\epsilon_{xy} = \frac{1}{2}\left(\frac{\partial u}{\partial y} + \frac{\partial v}{\partial x}\right)$$

The displacement gradients $\partial u/\partial x$ and $\partial v/\partial y$ are obtained by differentiation of the u and v families of moiré fringes in a direction perpendicular to the lines of the respective reference array. The displacement gradients $\partial u/\partial y$ and $\partial v/\partial x$ are obtained by differentiation of the moiré fringe families above in a direction parallel to the lines of the respective reference array.

Displacement gradients are obtained by graphical, optical (mechanical), or numerical differentiation techniques. Graphical differentiation is done by plotting the moiré fringe order versus location along a desired axis and drawing graphically the tangent to this curve at a number of discrete points. Optical differentiation is obtained by superimposing a transparent replica of the fringe pattern over it and shifting it by a small amount in the direction of the desired differentiation. The resulting coarser moiré pattern (moiré of moiré or second order moiré), gives fringes which represent loci of approximately equal displacement gradient. Numerical differentiation is possible by inputting fringe order and location data into curve-fitting computer programs.

In the shadow moiré method the out-of-plane displacement is obtained from the relation

$$w = \frac{pn}{\tan\alpha} \tag{3.15}$$

where w is the out-of-plane displacement, p the array pitch, n the fringe order and α the angle between the light beam and the normal to the array.

3.2.5 Applications

3.2.5.1 Uniaxially Loaded Glass/Epoxy Plate With Hole

An application of conventional moiré techniques is illustrated in Fig. 3-11.[2] The specimen was a 25.4 cm × 66 cm (10 in. × 26 in.) panel of $[0/\pm45/0/\overline{90}]_s$ layup with a 2.5 cm (1 in.) diameter central hole. A 40 line/mm (1000 ℓpi) array of horizontal lines was bonded onto the

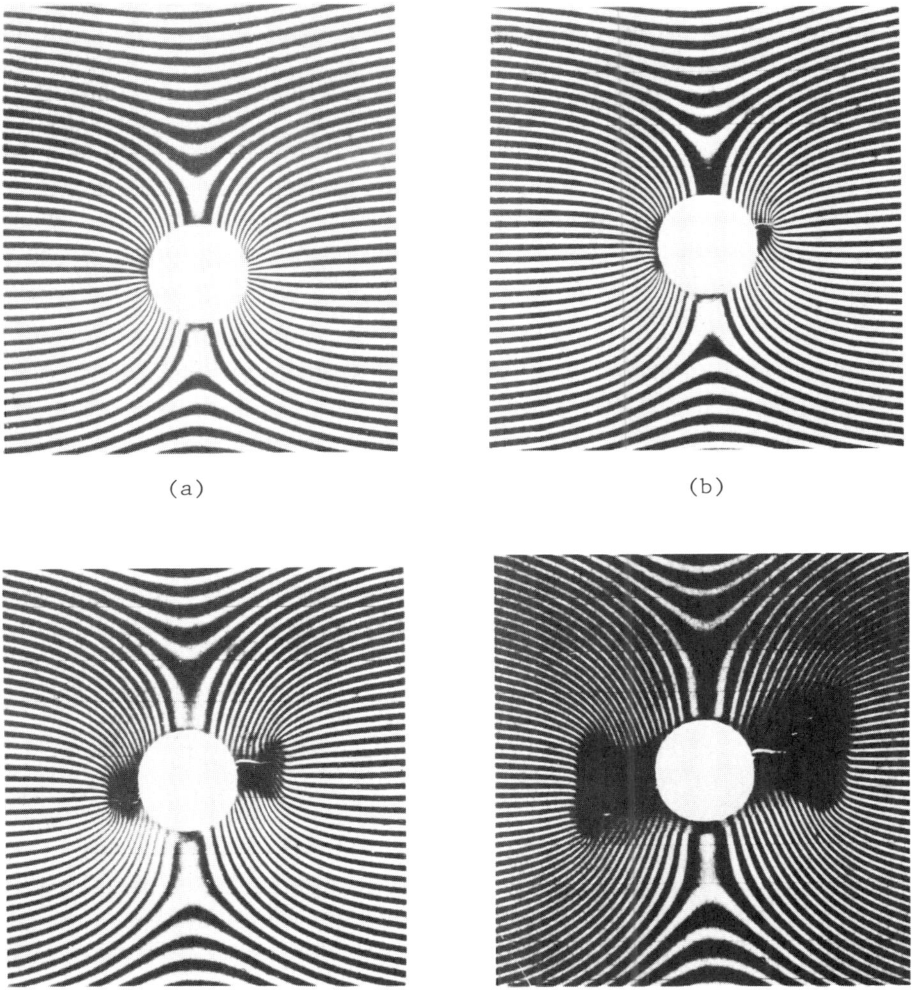

(a) (b) (c) (d)

Fig. 3-11—Sequence of moiré fringe patterns corresponding to vertical displacements in $[0/\pm45/0/\overline{90}]_s$ glass/epoxy specimen immediately prior to total failure. (a) p = 198 MPa (28,700 psi), (b), (c) p = 206 MPa (29,850 psi), (d) p = 210 MPa (30,400 psi)

specimen. It was loaded in uniaxial tension to failure. The tensile strain concentration factor determined from these moiré patterns is

$$(k_\epsilon)_{\theta=90\ deg} = 3.51$$

which compares well with the value of 3.50 determined by strain gages and photoelastic coatings on the same and similar specimens. Crack formation and propagation are evident at the higher stress levels. The cracks start on the boundary of the hole off the horizontal axis where the combination of stress and material strength becomes critical. Delamination, as evidenced by the dark regions in the moiré pattern, seems to follow crack propagation. One interesting phenomenon seen in the fringe pattern at the highest stress level is the seemingly discontinuous propagation of the cracks. The cracks seem to propagate in discrete, discontinuous horizontal steps with a gross direction of fracture at 45 deg. One possible explanation is the combination of two modes of failure, tensile fracture of the outer vertical fibers and shear delamination in the 45-deg interior plies.

The fringe multiplication techniques and procedures described before were applied to a similar glass/epoxy specimen. A grid of 20 dots/mm (500 dots/in.) was photoprinted on the specimen. The specimen was loaded in a testing machine in increments of load and the grid was photographed on high-resolution plates at each load level with the special camera shown in Fig. 3-10.

Replicas of the deformed grids were analyzed on an optical bench with a 200-ruling-per-millimeter analyzer grating for a tenfold multiplication. The reconstruction of fringe patterns was done according to procedures described by Post.[37] Two patterns were obtained from each grid replica, corresponding to vertical and horizontal displacements.

Figures 3-12 and 3-13 show moiré patterns for vertical and horizontal displacements (v- and u-displacement fields, respectively). The moiré patterns were analyzed by graphical differentiation of plotted displacement curves. In two cases, at the two highest load levels, mechanical differentiation was also employed by forming a second order moiré pattern.[41,42] Matching positive and negative transparencies of the moiré patterns were obtained by contact printing. These patterns when superimposed produce a uniformly dark field, but when one is shifted in the y-direction a new coarser fringe pattern results. For a v-displacement moiré and a shift of Δy, the new fringes represent approximately contours of constant partial derivatives $\frac{\partial v}{\partial y}$.

Vertical strains along the horizontal axis (obtained by graphical differentiation) were first plotted as a function of load and found to vary linearly with load up to at least $p = 95$ MPa (13,800 psi). The strains at this load level, obtained by graphical and mechanical differentiation, are in excellent agreement with the strain distribution obtained by the linear, anisotropic elastic solution for an infinite homogeneous plate (Fig. 3-14).[43,44]

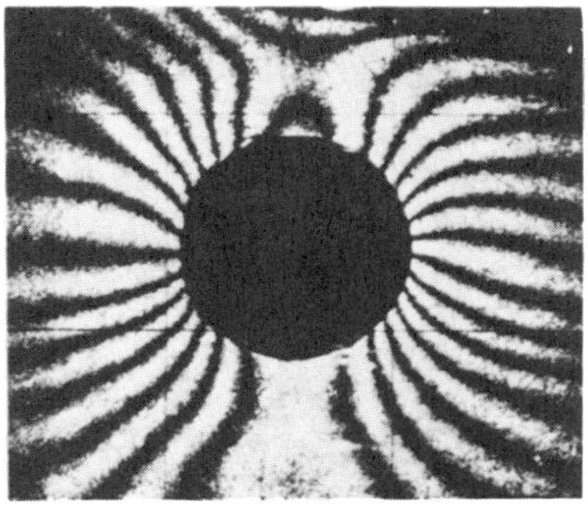

p = 3,450 psi

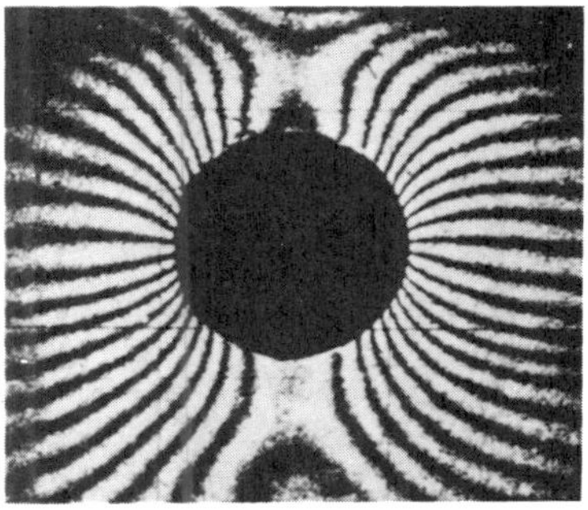

p = 5,170 psi

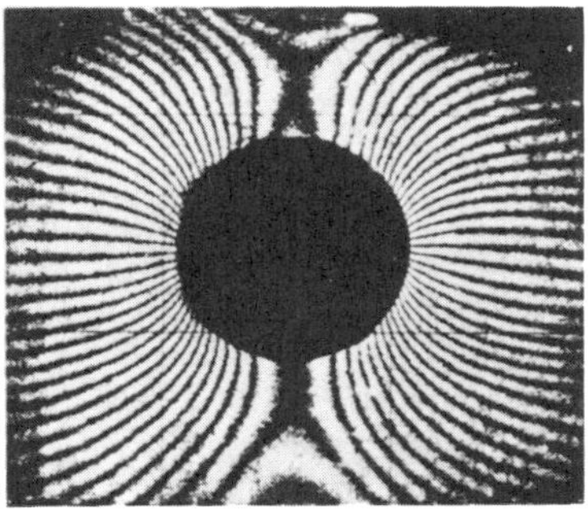

p = 8,600 psi

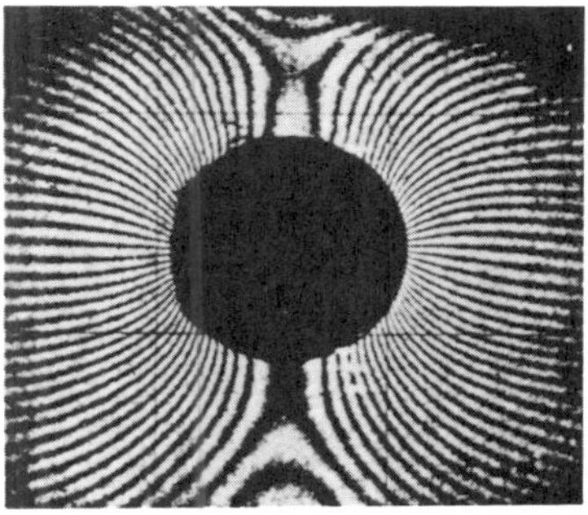

p = 10,300 psi

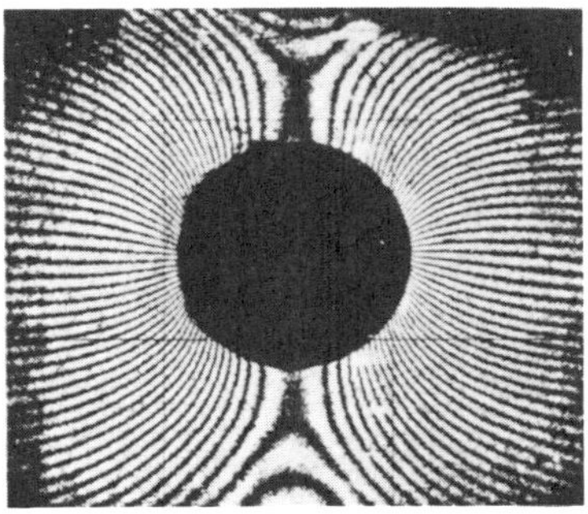

p = 11,950 psi

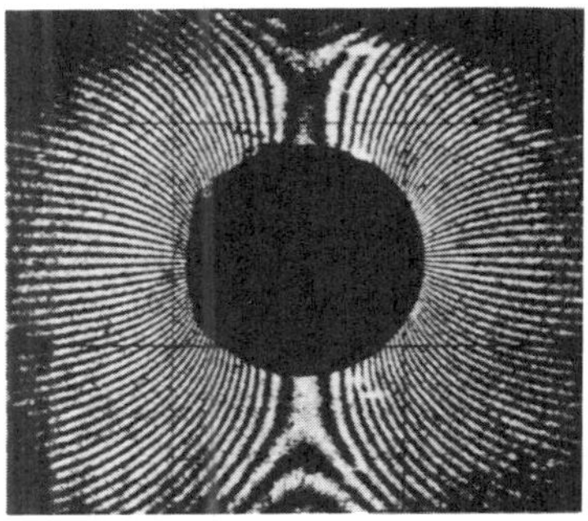

p = 13,800 psi

Fig. 3-12—Moire fringe patterns for vertical displacements around hole of glass/epoxy specimen $[0/\pm 45/0/\overline{90}]_s$ [fringe multiplication factor β = 10, corresponding to 200 lines/mm (5,000 lines per inch)]

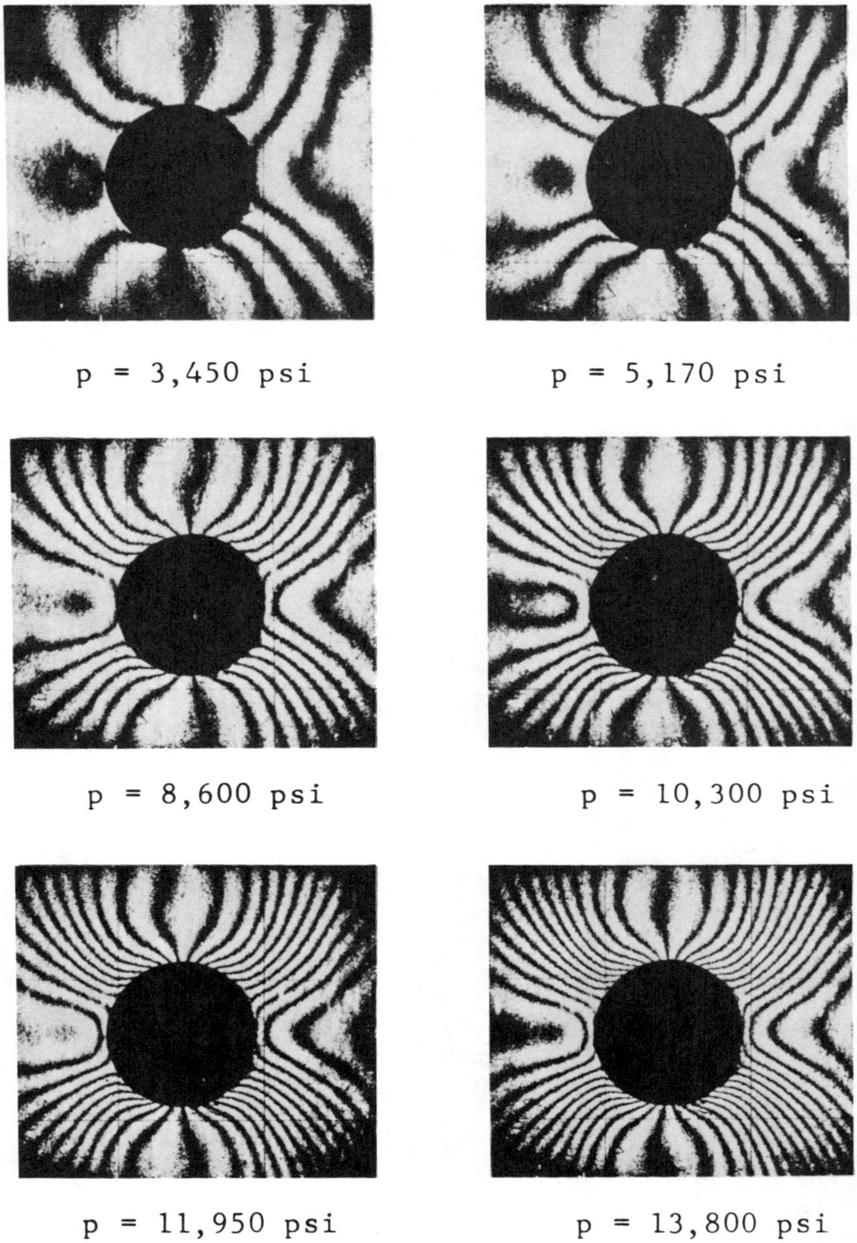

Fig. 3-13—Moiré fringe patterns for horizontal displacements around hole of glass/epoxy specimen $[0/\pm45/0/\overline{90}]_s$ [fringe multiplication factor $\beta = 10$, corresponding to 200 lines/mm (5,000 lines per inch)]

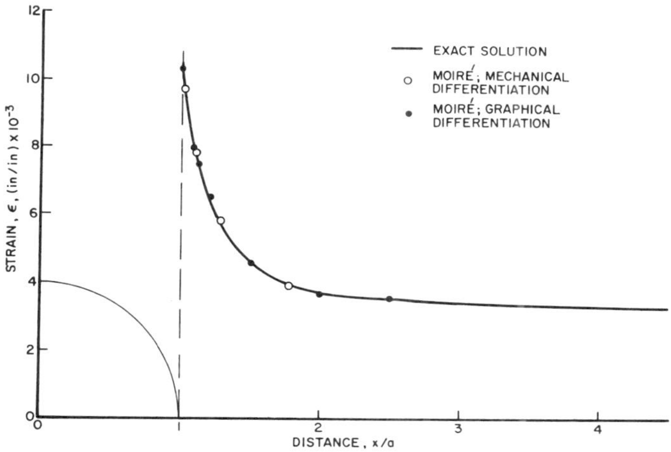

Fig. 3-14—Vertical strain distribution along horizontal axis in $[0/\pm 45/0/\overline{90}]_s$ glass/epoxy plate with hole at applied stress p = 95 MPa (13,800 psi)

3.2.5.2 Uniaxially Loaded Boron/Epoxy Plate with Hole

The same fringe-multiplication techniques above were applied to a 25.4 cm × 66 cm (10 in. × 26 in.) boron/epoxy panel of $[0/\pm 45/0/\overline{90}]_s$ construction with a 2.5 cm (1 in.) diameter hole.[3] The specimen was loaded in tension in increments of 15 MPa (2.2 ksi) up to an applied stress of 340 MPa (49.2 ksi). Replicas of the 500 dot-per-inch specimen grid were photographed with the moiré camera at each stress level up to 257 MPa (37.3 ksi). Fringe patterns were reconstructed with a 200 line-per-millimeter analyzer for a multiplication factor of β = 10.

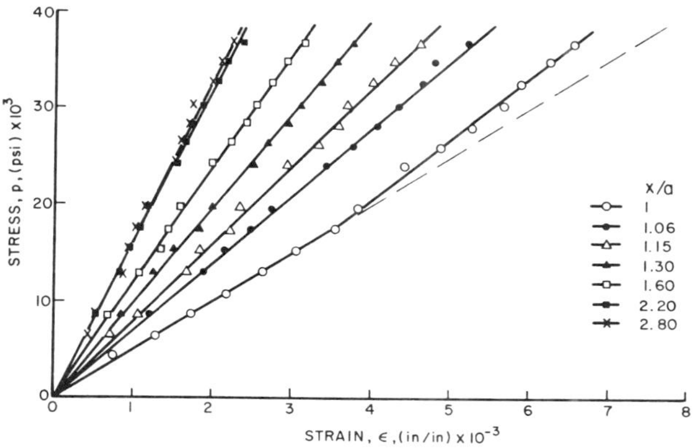

Fig. 3-15—Vertical strains along horizontal axis

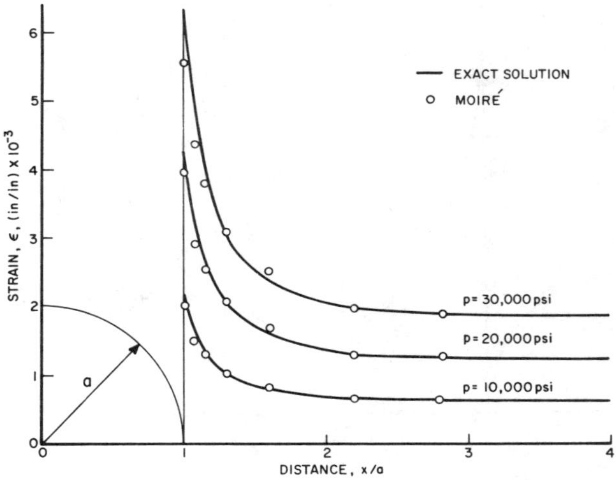

Fig. 3-16—Vertical strain distribution along horizontal axis at three stress levels

Vertical strains along the horizontal axis, obtained by graphical differentiation of the curves of fringe order versus location, are shown in Fig. 3-15. Away from the hole the strains are fairly linear with applied stress, but on the hole boundary a characteristic nonlinearity is shown at approximately p = 124 MPa (18,000 psi), which was also evident in results by other experimental methods.[3] A comparison of vertical strain distributions along the horizontal axis obtained by moiré and by the exact anisotropic elastic solution for a homogeneous infinite plate is shown in Fig. 3-16.[43,44] The agreement is good, except near the hole boundary where the experimental results are lower than the theoretical ones. This discrepancy increases with load. The horizontal strain distribution along the vertical axis of symmetry is shown in Fig. 3-17 for two levels of applied stress.

Tensile and compressive strain concentration factors for the points on the hole boundary on the horizontal and vertical axes, $\theta = 90$ deg and $\theta = 0$ deg, were computed for the linear range of strain response. Results from moiré data are:

$$(k_\epsilon)_{\theta = 90\ deg} = 3.33$$

$$(k_\epsilon)_{\theta = 0\ deg} = -1.49$$

compared to theoretical values of

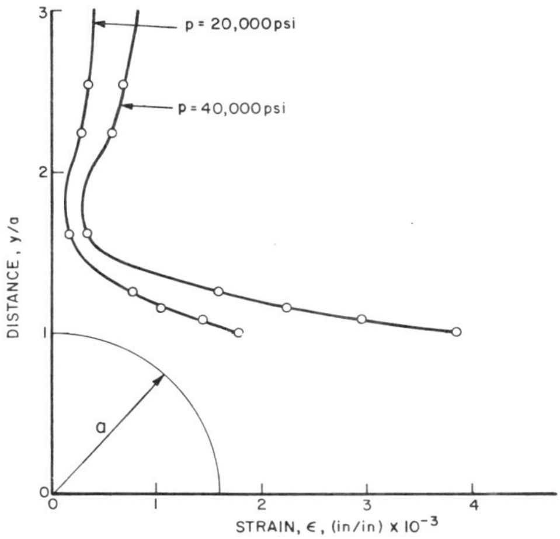

Fig. 3-17—Horizontal strain distribution along vertical axis at two applied stress levels

$$(k_\epsilon)_{\theta=90\ deg} = 3.45$$

$$(k_\epsilon)_{\theta=0\ deg} = -1.46$$

The fracture mechanics of this test have been discussed elsewhere.[3] A brief description of results is given here. Ultimate catastrophic failure was preceded by localized fractures which become audible at a fraction of the ultimate load. The applied stress at failure was approximately half the strength of a similar panel without a hole. The tensile strain on the horizontal axis reached values of over 8×10^{-3} prior to failure, approximately 20 percent higher than the unnotched coupon ultimate strain of 6.8×10^{-3}. It is conjectured that the capacity to support higher strains locally is due to the nonlinearity and nonuniformity of the strain distribution and to the steep gradient which confines the high strains and stresses to a small volume of the material. This is related to the notch size effect discussed in Section 2.8. The mode of failure was a combination of interlaminar shear and tensile cracking. Fracture initiated at points on the hole boundary located at 64 deg from the direction of loading, where the elastic shear stress and strain reached a maximum and where the tangential strain and shear strain increase nonlinearly with load and more rapidly than at any other point.

3.2.5.3 Glass/Epoxy Laminates with Cracks

An illustration of moiré patterns around a growing crack is given in Fig. 3-18. The specimen was a 10.2 cm (4 in.) wide glass/epoxy laminate

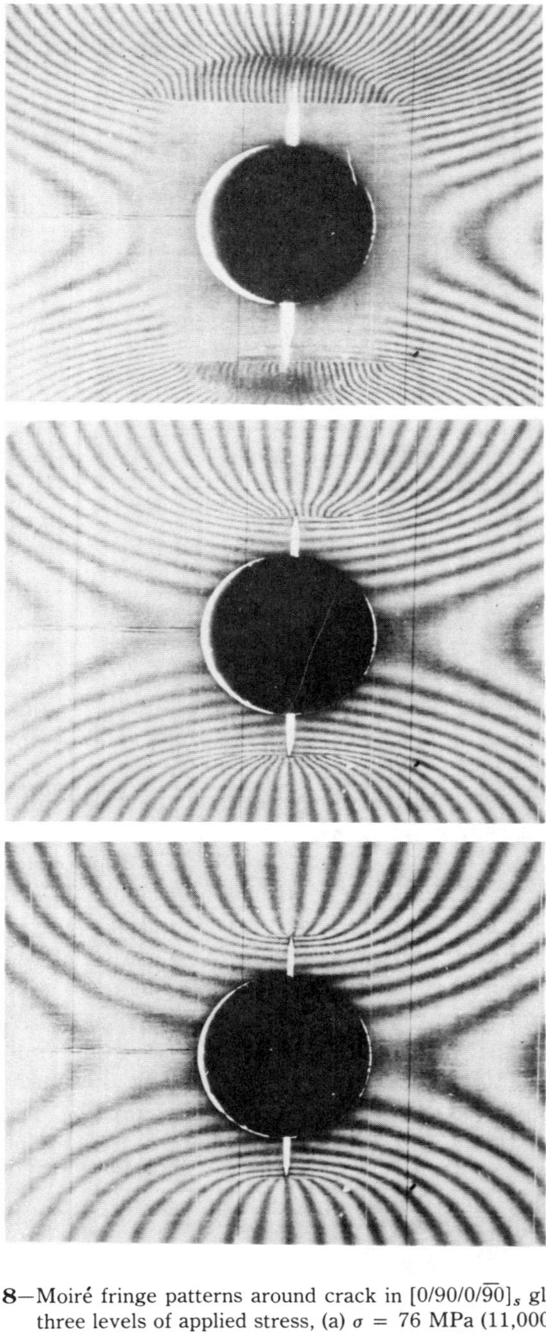

(c)

(b)

(a)

Fig. 3-18—Moiré fringe patterns around crack in $[0/90/0/\overline{90}]_s$ glass/epoxy composite at three levels of applied stress, (a) σ = 76 MPa (11,000 psi), (b) σ = 102 MPa (14,800 psi), (c) σ = 127 MPa (18,500 psi). (Ref. 38) of $[0/90/0/90]_s$ layup with a 2.5 cm (1 in.) diameter hole and two 6.3 mm (¼ in.) saw cuts on the hole boundary.[38] The fringe patterns represent vertical displacements at three load levels. These patterns display certain characteristics related to the mode of failure of cross-ply laminates. The discontinuity of the fringe pattern along the vertical line tangent to the crack tip is a manifestation of surface failure. This line represents a boundary between high tensile strain in front of the crack and high shear strain behind the tip of the crack. This is corroborated by the fact that the fringes change abruptly from nearly horizontal in front of the crack to nearly vertical behind it. The high inplane shear stress near the crack tip causes separation of the vertical 0-deg fibers. This failure is accompanied by subsurface delamination and separation of the 90-deg fibers with the crack propagating along the subsurface 90-deg ply. The delamination extends over two circular segments, concentric with the hole, ahead of the crack. The surface 0-deg fibers that have separated from each other and from the 90-deg ply below behave as a bundle of fibers, hence they are under nearly uniform strain in the segment-shaped area. This strain is approximately five times the far-field vertical strain. As the load is increased, the original vertical crack grows in length and new vertical cracks appear at the tip of the propagating horizontal crack.

Another application of moiré to crack measurement in composite materials was discussed by Chiang and Slepetz.[45] An example of a propagating crack in a double cantilever beam of unidirectional glass/epoxy is shown in Fig. 3-19. An array of 12 lines/mm (300 lines/in.) parallel to the fiber direction and the propagating crack was applied. In Fig. 3-19a the reference ruling was rotated slightly with respect to the model ruling, thus producing the inclined fringes. In Fig. 3-19b both model and reference rulings are horizontal, but of slightly different pitch. The fringe patterns around the crack show "kinks" in the former case and bifurcations in the latter case because the displacement field is not single-valued around the crack.

3.2.5.4 Uniaxially Loaded Graphite/Epoxy Plate with Crack

Conventional moiré techniques were used to determine displacements and strains in a $[0/\pm45/90]_s$ graphite/epoxy laminate with a crack.[8] The specimen was 12.7 cm (5 in.) wide and 56 cm (22 in.) long with a 1.27 cm (0.50 in.) through-the-thickness horizontal crack. An array of 40 lines/mm (1000 lines/in.) was applied to the specimen surface around the crack. The specimen surface was made reflective by depositing a thin coating of epoxy dyed white with titanium dioxide. The film with the line array was then bonded on this white reflective surface.

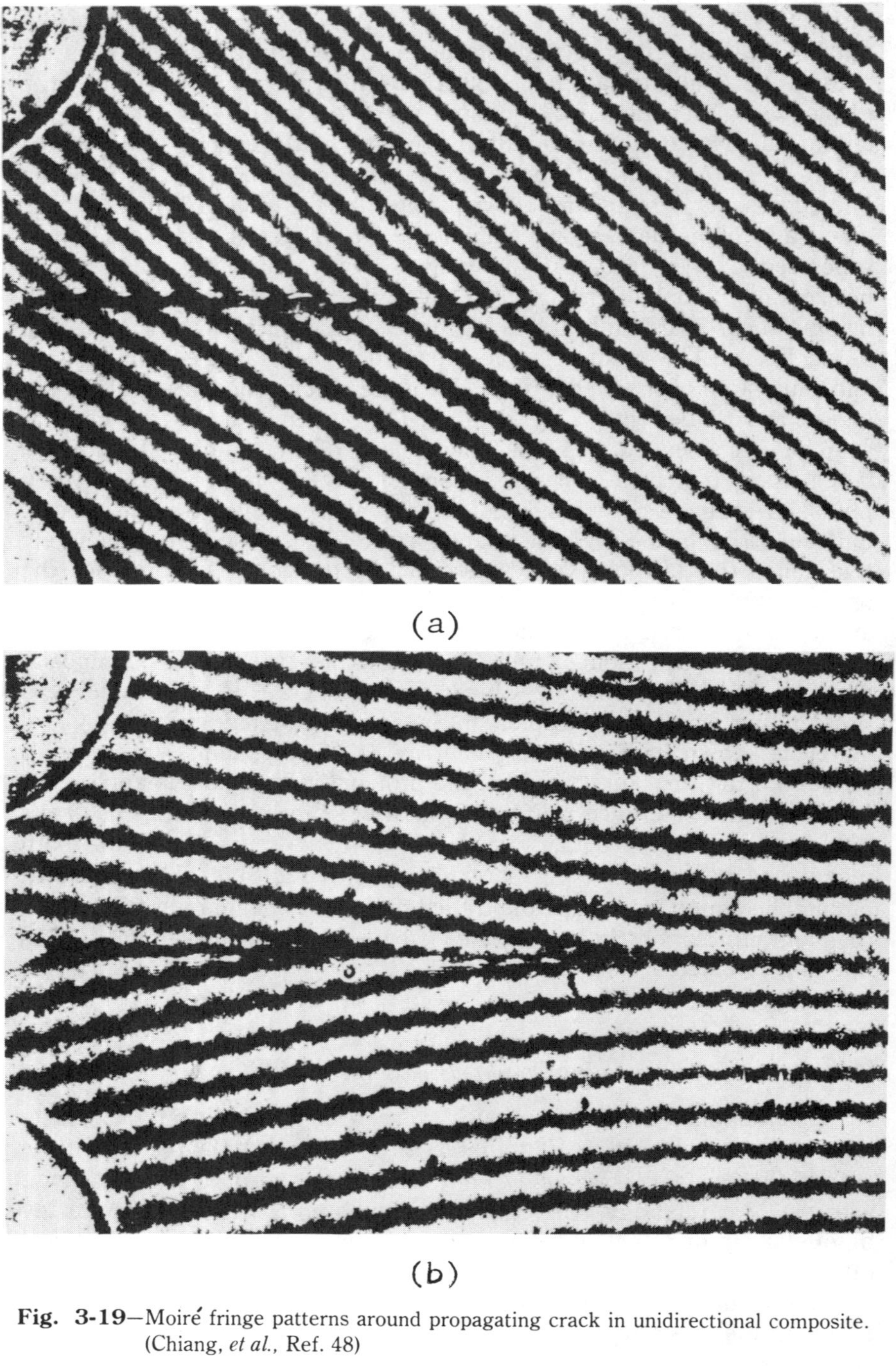

(a)

(b)

Fig. 3-19—Moiré fringe patterns around propagating crack in unidirectional composite. (Chiang, *et al.*, Ref. 48)

(c)

(b)

(a)

Fig. 3-20—Moiré fringe patterns around crack in uniaxially loaded $[0/\pm 45/90]_s$ graphite/epoxy specimen for three levels of applied stress: (a) σ_{xx} = 152 MPa (22 ksi), (b) σ_{xx} = 202 MPa (29 ksi), (c) σ_{xx} = 253 MPa (37 ksi)

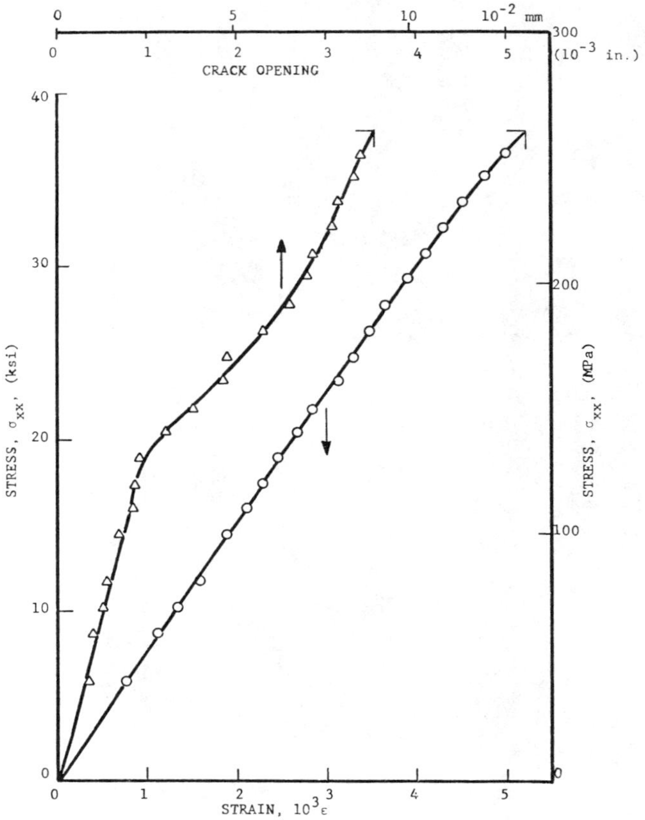

Fig. 3-21—Crack opening displacement and far-field strain for $[0/\pm 45/90]_s$ graphite/epoxy specimen with a 1.27 cm (0.50 in.) horizontal crack

Moiré fringe patterns corresponding to vertical displacements in the vicinity of the crack are shown in Fig. 3-20 for three levels of load. Each fringe represents a locus of points of constant vertical displacement of 0.025 mm (0.001 in.) relative to its neighboring fringe. These fringe patterns were analyzed to yield far-field strains at a distance of $\frac{y}{a} \simeq 3$ and the crack opening displacement at the center of the crack (Fig. 3-21). The plot for the latter indicates that the crack opening displacement becomes nonlinear and increases at an increasing rate at an applied stress of approximately 138 MPa (20 ksi) which is near the level of rapid strain increase near the crack tip obtained with strain gages near the crack tip. Final failure was preceded by cracking noises heard at a stress level of 172 MPa (25 ksi).

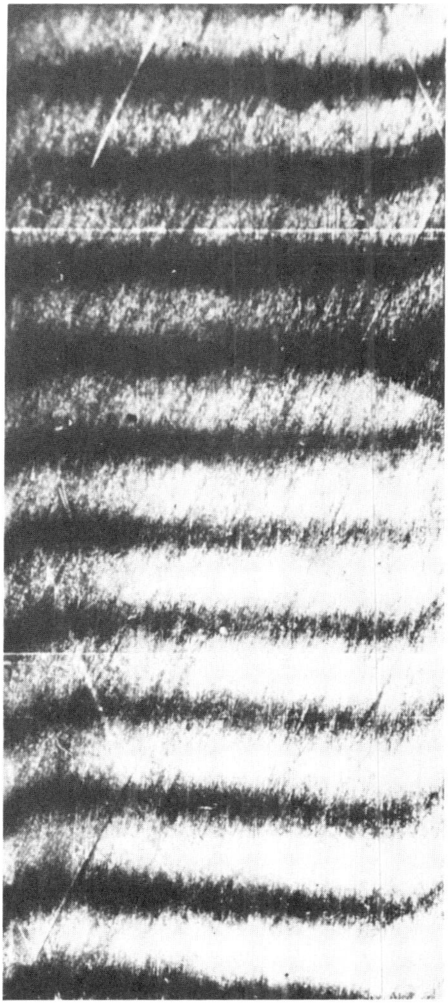

Fig. 3-22—Moiré fringe pattern in axially loaded $[25_4/-25_4]_s$ graphite/epoxy specimen. (Ref. 46)

3.2.5.5 Edge Effects in Angle-Ply Laminates

The interlaminar shear effect near a free edge was studied by Pipes and Daniel.[46] Fringe patterns corresponding to axial displacements in a $[+25_4/-25_4]_s$ graphite/epoxy specimen under axial tension are shown in Fig. 3-22. The fringes have a characteristic *S*-shape with pronounced curvature near the edges. The moiré results were in agreement with

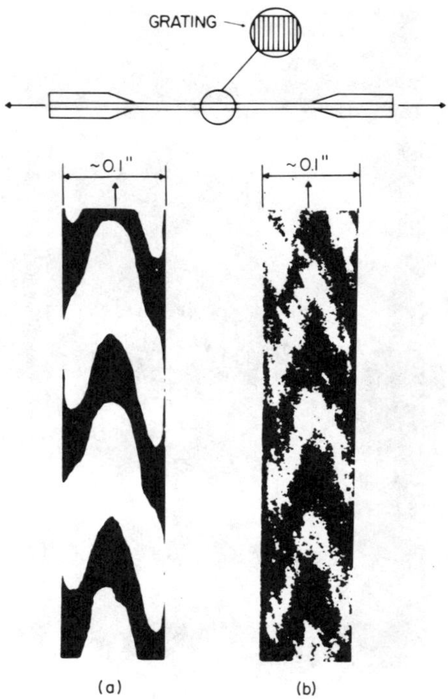

Fig. 3-23—Moiré fringe pattern showing edge effect in angle-ply composite laminate. (a) 4 lines/mm, (b) 100 lines/mm. (Chiang, *et al.*, Ref. 48)

theoretical results by Pipes and Pagano[47] as shown in Fig. 2-19. A similar effect of interlaminar shear on the edge of an angle-ply specimen was illustrated by Chiang *et al.* (Fig. 3-23).[48]

3.2.5.6 Applications of Shadow Moiré

Shadow moiré techniques have been applied to the study of buckling behavior of composite panels.[49,50] Unstiffened and stiffened graphite/epoxy panels were loaded under inplane compression and shear.[49] Buckling and postbuckling behavior were clearly illustrated by shadow moiré patterns. Load deflection characteristics obtained experimentally were in reasonable agreement with theoretical predictions.

3.2.6 Discussion

Moiré techniques are well suited for determining full-field inplane displacement fields in composite laminates. These displacement fields are primarily used to determine strain distributions and strain concentrations around geometrical discontinuities, such as cracks and holes. In many cases, they even elucidate the initiation, propagation, and mode of failure. Conventional techniques, using specimen rulings of 40 lines/mm (1000 lines/in.), seem to yield satisfactory results in glass/epoxy laminates and in boron/epoxy composites with predominantly ±45-deg plies. No measurable differences have been observed between specimens with the rulings bonded and those with rulings photoprinted on the specimen surface. In the case of opaque laminates, the only additional preparation required is the vacuum-deposition of an aluminum coating or white paint to render the surface reflective.

Sensitivity enhancement through fringe-multiplication techniques is always desirable and in many applications to composites it is necessary. Surface smoothness is very important. Deposition of a thin epoxy coating has proven satisfactory for tenfold multiplication. Projection photography of the specimen grid requires the type of distortion-free special purpose camera described here. This is not the only means of recording the deformed specimen grids. Alternate means, such as contact photography of electro-luminescent specimen grids or reflex contact photography, could be developed. One limitation of the technique used here is the limited field of the camera: 6.4 cm × 7.6 cm (2½ in. × 3 in.). To cover a large area, this would require special mountings for the camera to allow motion parallel to the surface of the specimen.

With the techniques discussed here the upper practical limit seems to be tenfold multiplication; however, these techniques could be refined for even higher multiplication. Mechanical differentiation by second order moiré seems to be a quick means of obtaining partial derivatives; however, it can be used only with sufficiently dense fringe patterns. These patterns would be obtained at higher load levels and with higher multiplication factors.

3.3 BIREFRINGENT COATINGS

3.3.1 Theory of Birefringent Coatings

Birefringent or photoelastic coatings have been applied successfully to isotropic materials for several years.[51-54] The method consists of bonding a thin sheet of photoelastic material to the surface of the specimen, such that the bonded interface is reflective. When the specimen is loaded, the surface strains are transmitted to the coating and produce a fringe pattern which is recorded and analyzed by means of a reflection polariscope. For perfect strain transmission

$$\epsilon_1^c = \epsilon_1^s$$
$$\epsilon_2^c = \epsilon_2^s \tag{3.16}$$

where the superscripts c and s refer to the coating and specimen, respectively. The difference in principal strains and the coating birefringence are related by the strain-optic law:

$$\epsilon_1^c - \epsilon_2^c = \epsilon_1^s - \epsilon_2^s = \frac{Nf_\epsilon}{2h} = NF_\epsilon \tag{3.17}$$

where f_ϵ is the strain fringe value, F_ϵ the coating fringe value, h the coating thickness, and N the fringe order.

The application of coatings to composite materials is based on the same principles. The surface strain field of the anisotropic composite produces a photoelastic response in the isotropic coating. The birefringence in the coating is related to the difference in principal strains by the strain-optic law above. Techniques, such as the oblique incidence technique, exist for determining the individual values of the principal strains.[55] The stresses in the composite are computed from the strains using the appropriate anisotropic stress–strain relations.

When the composite material is orthotropic with the elastic axes of symmetry 1, 2 coinciding with the axes of material, geometric and loading symmetry, the analysis is simplified. Then, the coating birefringence is related to the principal stresses in the composite by

$$\frac{Nf_\epsilon}{2h} = \frac{\sigma_{11}^s}{E_{11}}(1 + \nu_{12}) - \frac{\sigma_{22}^s}{E_{22}}(1 + \nu_{21}) \tag{3.18}$$

where E_{11}, E_{22}, ν_{12}, ν_{21} are the elastic constants referred to the axes of elastic symmetry.

3.3.2 Limitations Near Free Edge

Applications of birefringent coatings to glass/epoxy composites were described by Dally and Alfirevich.[56] Similar applications to glass/, boron/ and graphite/epoxy composites were described by Daniel, Rowlands and Whiteside[2,3,5-11,57] and by Yeow, Morris and Brinson.[58] A critical discussion of the method was given by Pipes and Dalley, who conducted detailed analyses near free edges of composite materials and pointed out some limitations of the coating method.[59] One of the major limitations is the effect of Poisson's ratio mismatch.[53] If Poisson's ratio of the coating is different from that of the substrate a distortion in the displacement field is produced in the coating which is especially pronounced at the boundaries.

For a specially orthotropic laminate, it can be assumed that the maximum principal strain is parallel to the boundary and is transmitted to the coating ($\epsilon_1^c = \epsilon_1^s$) and that the minimum principal strain is given by $\epsilon_{22}^s = -\nu_{12}^s \epsilon_{11}^s$. The transverse strain in the coating at the boundary varies between

$$\epsilon_{22}^c = \epsilon_{22}^s = -\nu_{12}^s \epsilon_{11}^s \tag{3.19}$$

at the interface, and

$$\epsilon_{22}^c = -\nu^c \epsilon_{11}^c = -\nu^c \epsilon_{11}^s \tag{3.20}$$

at the free surface. It has been shown experimentally that eq (3.20) above represents fairly accurately the average transverse strain in the coating.[56] Thus, the principal strain along the boundary can be obtained from the relation

$$\epsilon_{11}^s = \frac{Nf_\epsilon}{2h} \frac{1}{1 + \nu^c} \tag{3.21}$$

and the nonzero principal stress along the boundary is

$$\sigma_{11}^s = \frac{E_{11}}{1 + \nu^c} \frac{Nf_\epsilon}{2h} \tag{3.22}$$

Difficulties arise also in a small transition region near the boundary, where the photoelastic response is a function of both Poisson's ratios, those of the coating and the composite. For a uniaxial state of stress the strain can be expressed as[56]

$$\epsilon_{11}^s = \frac{Nf_\epsilon}{2h} \frac{1}{1 + \nu_{12}^s + C_\nu(\nu^c - \nu^s)} \tag{3.23}$$

where C_ν is a correction factor ranging in value from 1 at the boundary to zero away from it. The transition region extends for approximately four coating thicknesses from the boundary. This transition region, of course, is reduced by decreasing the coating thickness and is completely eliminated when Poisson's ratio of the composite equals that of the coating.

The discussion above pertaining to the photoelastic response near a free boundary is valid only for a specially orthotropic laminate with elastic, material, and principal strain axes parallel and normal to the boundary. These assumptions are not always valid, since, as Pipes and

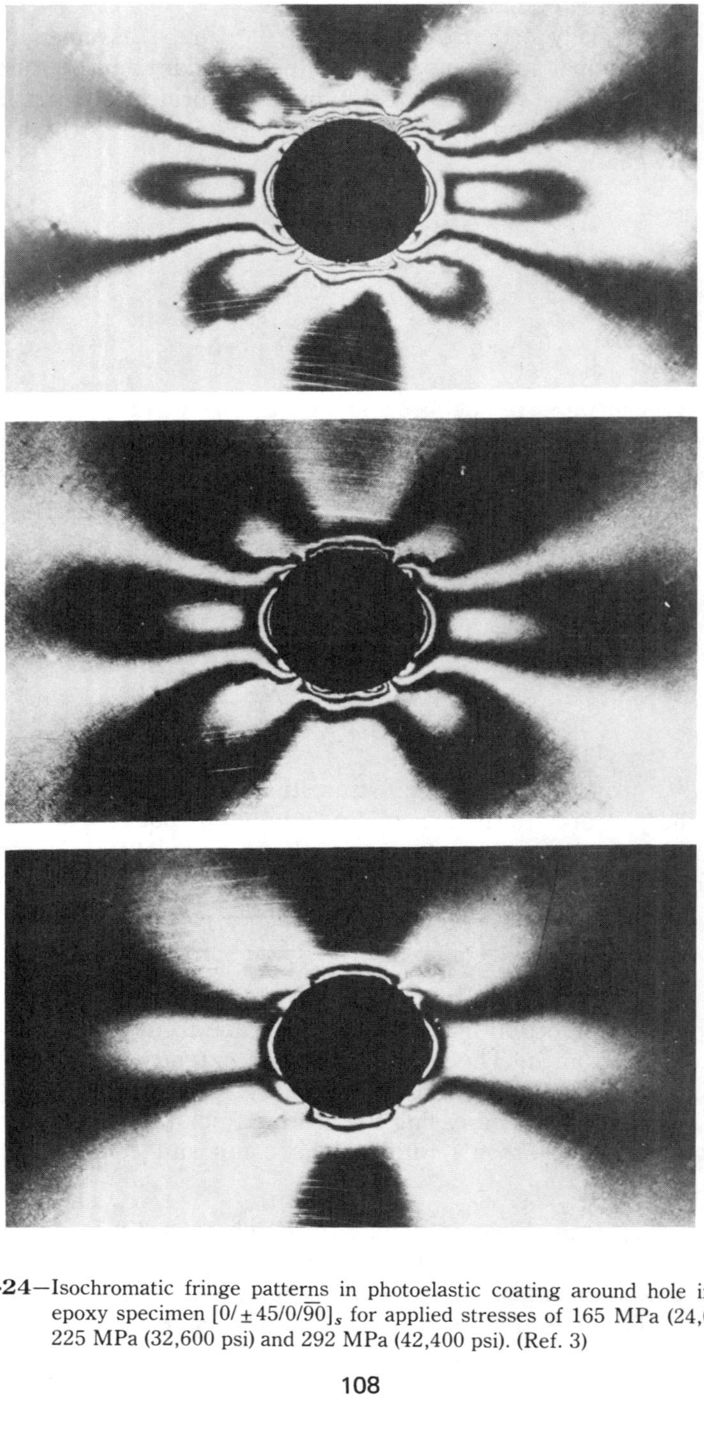

Fig. 3-24—Isochromatic fringe patterns in photoelastic coating around hole in boron/epoxy specimen $[0/\pm 45/0/\overline{90}]_s$ for applied stresses of 165 MPa (24,000 psi), 225 MPa (32,600 psi) and 292 MPa (42,400 psi). (Ref. 3)

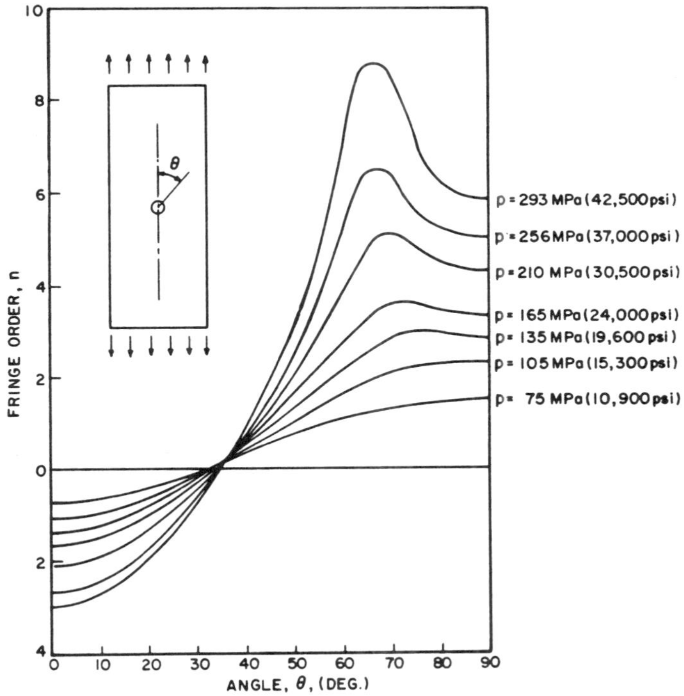

Fig. 3-25—Birefringence distribution around boundary of hole for various stress levels. (Ref. 3)

Dalley[59] pointed out, appreciable shear deformation may exist near a free boundary, especially when the fibers of the surface plies are not parallel or normal to it. Then, the principal strains near the free boundary are neither parallel nor normal to it. Depending on the fiber orientations, the transverse strain at the boundary can vary appreciably through the thickness of the laminate and, thus, the birefringence in the coating may reflect primarily the characteristics of the top ply. These effects are reduced appreciably in thin laminates with interspersed fibers, when the top ply has fibers parallel or normal to the boundary, and in cases where the fiber is large compared to the ply thickness.

3.3.3 Applications

3.3.3.1 Boron/Epoxy Plate with Hole

Birefringent coatings have been used to study the deformation and failure around a circular hole in a boron/epoxy composite plate under

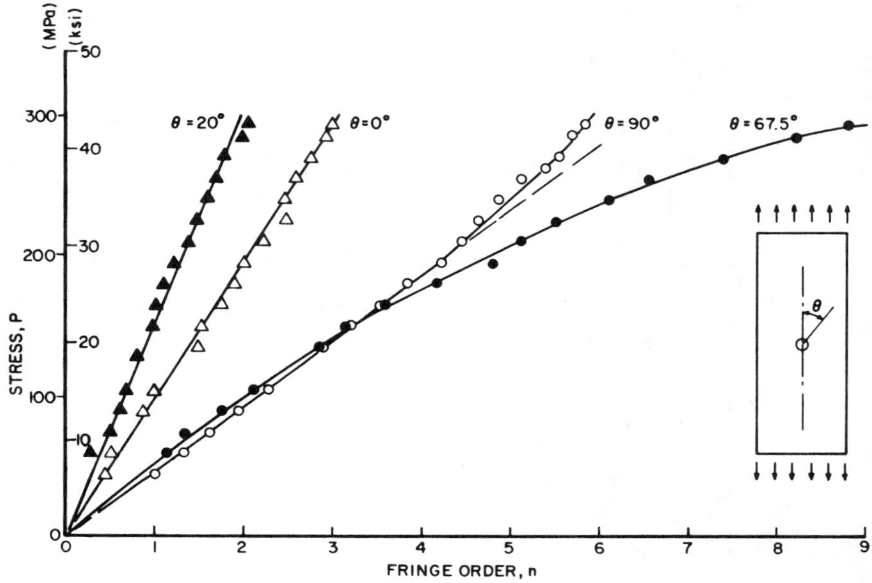

Fig. 3-26—Birefringence in coating on the boundary of the hole as a function of applied stress for $[0/\pm45/0/\overline{90}]_s$ boron/epoxy specimen. (Ref. 3)

uniaxial tension.[3] The specimen was a 25.4 cm × 66 cm (10 in. × 26 in.) boron/epoxy panel of $[0/\pm45/0/\overline{90}]_s$ layup with a 2.54 cm (1 in.) diameter hole. A 1 mm (0.04 in.) thick coating was applied. Figure 3-24 shows isochromatic fringe patterns on the coating for three levels of applied average stress, 165 MPa (24,000 psi), 225 MPa (32,600 psi), and 292 MPa (42,400 psi). The birefringence (proportional to strain) variation around the boundary of the hole for various load levels is shown in Fig. 3-25. An important and significant result is that the location of maximum birefringence shifts from $\theta = 90$ deg (horizontal axis) at low loads to $\theta = 66$ deg near the failure load. A plot of birefringence as a function of load for various locations on the hole boundary (Fig. 3-26) shows that it varies nearly linearly at $\theta = 0$ deg and $\theta = 20$ deg. At $\theta = 90$ deg, the birefringence is linear up to an applied average stress of 195 MPa (28,300 psi), and thereafter it increases at a lower rate. The birefringence at $\theta = 67.5$ deg, near the point of maximum birefringence at higher loads, is linear up to approximately $p = 97$ MPa (14,000 psi), and thereafter it varies with load at an increasing rate.

The strain distribution around the hole boundary at an applied stress level of $p = 69$ MPa (10,000 psi) was computed from the photo-

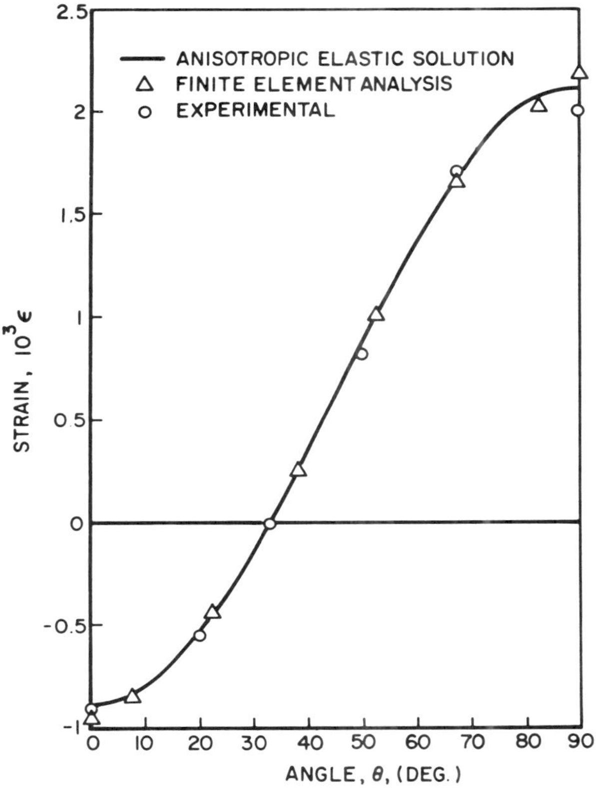

Fig. 3-27—Tangential strain distribution around boundary of hole in $[0/\pm 45/0/\overline{90}]_s$ boron/epoxy plate for p = 69 MPa (10,000 psi) applied stress. (Ref. 3)

elastic data using the strain-optic relation of eq (3.21) and compared in Fig. 3-27 with results of a finite element analysis and the anisotropic elastic solution.[43,44] The agreement between the photoelastic coating and the other results is very good except at the points of highest tensile and compressive strain. The isotropic point, point of zero tangential normal stress and strain, occurs at θ = 33 deg according to theoretical and experimental results.

Tensile and compressive strain concentration factors computed from the photoelastic data within the linear range are

$$(k_\epsilon)_{\theta=90\ deg} = 3.09$$

$$(k_\epsilon)_{\theta=0\ deg} = -1.46$$

Fig. 3-28—Failure around hole in $[0/\pm 45/0/\overline{90}]_s$ boron/epoxy panel under tension. (Ref. 3)

which compare very well with the corresponding values of 3.08 and −1.42 determined from strain gage data.

The failure process was clearly manifested in the isochromatic fringe patterns of Fig. 3-24. The birefringence distributions (Figs. 3-25 and 3-26) show clearly how the fringe order, hence tangential normal strain, increases rapidly in a nonlinear fashion at around $\theta = 67.5$ deg. This is related to the fact that the shear stress-shear strain (σ_{xy} versus ϵ_{xy}) and circumferential stress–strain ($\sigma_{\theta\theta}$ versus $\epsilon_{\theta\theta}$) response of the material at $\theta = 67.5$ deg is highly nonlinear. Failure initiated near the points of maximum birefringence. At these points, the elastic membrane and interlaminar shear stresses reach maximum values, and the circumferential normal strain and shear strain increase nonlinearly with load and more rapidly than at any other point. A complex failure combining tensile cracking and shear delamination modes produced a lip on the hole boundary. The crack then progressed parallel to the horizontal axis to complete failure (Fig. 3-28).

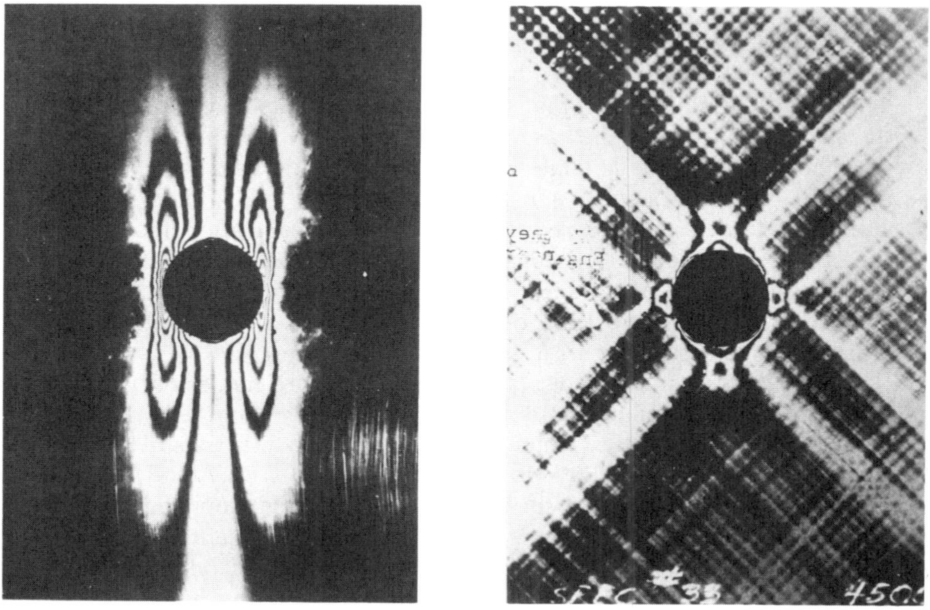

(a) (b)

Fig. 3-29—Isochromatic fringe patterns in photoelastic coating around hole in boron/epoxy specimens. (a) $[0/90/0/90]_s$, p = 169 MPa (24,600 psi), (b) $[\pm 45/\pm 45]_s$, p = 77 MPa (11,100 psi). (Ref. 5)

3.3.3.2 Influence of Laminate Construction

The influence of laminate construction on the behavior of boron/epoxy tensile plates with holes has also been studied with photoelastic coatings.[5] Some of the laminate constructions studied, in addition to the one discussed above, were $[0/90/0/90]_s$, $[45/90/0/-45]_s$, $[\pm 45/0/\pm 45]_s$ and $[\pm 45/\pm 45]_s$. The first and last of these represent two extremes of behavior illustrated by the fringe patterns of Fig. 3-29. The $[0/90]_{2s}$ construction is characterized by a low maximum strain at failure (6.6×10^{-3}) contrasted with very large strains ($>25 \times 10^{-3}$) at failure for the $[\pm 45]_{2s}$ construction. Another apparent difference is the sharp strain gradient and high strain concentration (4.82) in the $[0/90]_{2s}$ construction contrasted with a moderate gradient and low strain concentration (2.06) in the $[\pm 45]_{2s}$ layup. In the former case the fringes are concentrated near the hole around the horizontal axis and a uniform far-field birefringence is attained at a distance of between half and one radius from the boundary of the hole. In the case of the $[\pm 45]_{2s}$ specimen the influence of the hole extends along the 45-deg radii through the entire width of the plate. This is a result of the heterogeneous nature of the composite material in which the boron fibers transfer the disturbance along their length. It is an illustration that the St. Venant principle cannot apply in the same manner as in the case of a homogeneous, isotropic material.

The influence of laminate construction on strength was very pronounced. Laminates with a high percentage of 0-deg plies but with sufficient 45-deg plies to reduce the stress concentration factor were the strongest. The $[0/90]_{2s}$ construction with 50 percent 0-deg plies is not strong because of the high stress concentration factor. The $[\pm 45]_{2s}$ construction is the weakest because of the absence of 0-deg plies, although the stress concentration factor is the lowest.

The various laminates displayed different failure modes as expected. Figure 3-30 shows failure patterns of four of these. The $[0/90]_{2s}$ laminate failed in the most brittle fashion through the horizontal axis. The quasi-isotropic laminate $[45/90/0/-45]_s$ failed like the $[0/\pm 45/0/\overline{90}]_s$ one discussed earlier, i.e., failure originated on the hole boundary off the horizontal axis, progressed at some angle to the horizontal and finally propagated across the width of the plate in a horizontal direction. Failure in the $[\pm 45/0/\pm 45]_s$ laminate originated off the horizontal axis on the hole boundary and propagated across the width at some angle to the horizontal. The $[\pm 45]_{2s}$ laminate failed primarily in shear at 45-deg to the horizontal.

3.3.3.3 *Influence of Stacking Sequence*

The influence of stacking sequence on laminate strength has been observed and discussed by several investigators.[5,60,61] Pagano and Pipes discussed published experimental results and proposed a theoretical explanation of observed variations in strength with stacking sequence.[60] The observed differences in laminates of the same construction but varying stacking sequence are attributed to interlaminar stresses near free edges.* The most influential of these stresses is the interlaminar normal stress. When a laminate is subjected to an axial stress, the differences in Poisson's ratio between the various plies result in transverse stresses. These stresses are equilibrated by interlaminar shear stresses and couples produced by the interlaminar normal stresses. When the outer plies are in transverse tension, the interlaminar normal stresses near the free boundary are tensile, tending to delaminate the composite and, therefore, decrease its strength. The opposite is true when the outer plies are in transverse compression. Stacking-sequence variations also introduce variations in residual thermal stresses. If the

---

**For a detailed discussion of free-edge effects see Section 2-9.*

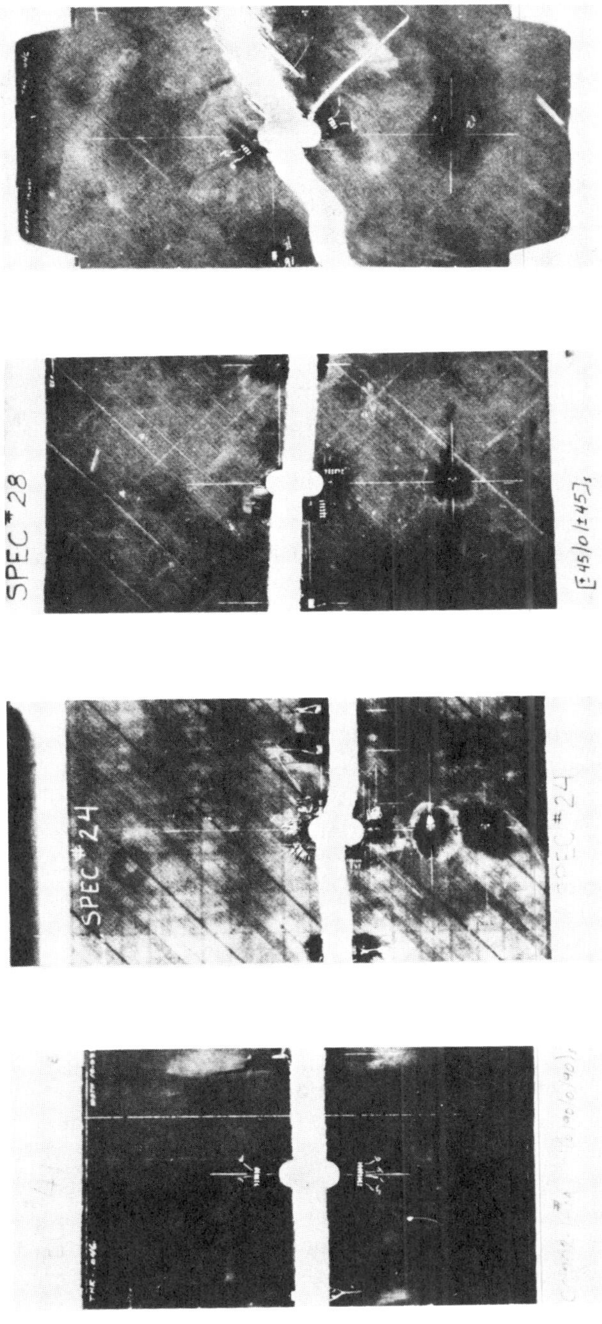

$[\pm 45/\pm 45]_s$

$[\pm 45/0/\pm 45]_s$

$[45/90/0/-45]_s$

$[0/90/0/90]_s$

Fig. 3-30—Failure patterns of boron/epoxy panels with holes of various laminate constructions. (Ref. 5) free boundary contains a region of high stress concentration (hole) forcing initiation of failure in that region, the effects of stacking-sequence variation on strength are further accentuated. The behavior of a laminate at a free boundary depends on many factors, including the membrane stress parallel to the boundary, the interlaminar shear and normal stresses, residual thermal stresses, and geometry of the boundary.

The analysis referred to above pertained to a free straight edge in a uniaxially loaded laminate. Rybicki and Hopper[61] extended it to a case of a free circular boundary. Although the solutions for the two cases are fundamentally different, they tend to approach each other as the hole radius becomes large compared to the laminate thickness. The analyses above are linear and assume each ply to be homogeneous. The nonlinearity of response near failure and the heterogeneity of the ply coupled with the very steep gradient of the interlaminar normal stress make strength predictions on the basis of interlaminar stresses qualitative only.

Pronounced differences were observed in a pair of $[0_2/\pm 45/\bar{0}]_s$ and $[\pm 45/0_2/\bar{0}]_s$ layups. The first layup with a measured strain-concentration factor of 3.58 failed at 498 MPa (72,200 psi) compared to 425 MPa (61,700 psi) for its stacking-sequence variation $[\pm 45/0_2/\bar{0}]_s$ with a measured strain-concentration factor of 4.02. The development of interlaminar stresses near the boundary is illustrated in Fig. 3-31. The laminate Poisson's ratio of 0.61 is much higher than that of the 0-deg outer plies in the first laminate, thus producing transverse compression in these plies, which in turn results in beneficial compressive interlaminar normal stresses near the boundary. The situation is reversed for the $[\pm 45/0_2/\bar{0}]_s$ layup, thus resulting in decreased strength for the latter. By taking successive free-body diagrams with additional plies like those shown in Fig. 3-31, the complete distributions of transverse normal stress σ_z are obtained.

Some of the differences between the two layups mentioned are manifested in the isochromatic-fringe patterns in the photoelastic coating around the hole (Fig. 3-32). The pattern for the first layup is fairly symmetrical, with lower birefringence, and shows a tendency for failure propagation along the vertical tangents to the hole boundary, i.e., parallel to the 0-deg outer plies. The second pattern is antisymmetrical, with higher birefringence concentration. The failures in the first and third quadrants are associated with shear failure of the outer ±45-deg plies, whereas the failures in the second and fourth quadrants are associated with crack propagation along the subsurface 0-deg plies. The peak birefringence, an indication of impending failure, is much more pronounced in the $[\pm 45/0_2/\bar{0}]_s$ layup. The influence of interlaminar stresses near the boundary is difficult to assess due to shifting of the

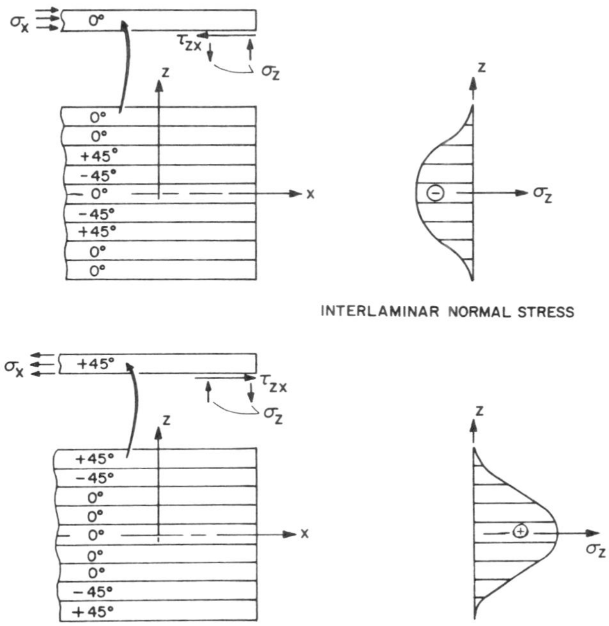

Fig. 3-31—Interlaminar stresses in outer ply and distribution of interlaminar normal stress near the boundary for $[0_2/\pm 45/\bar{0}]_s$ and $[\pm 45/0_2/\bar{0}]_s$ laminates under unaxial tensile loading

point of maximum strain and the nonlinearity of the response at points off the horizontal axis.

One dramatic effect of the stacking-sequence variation of the pair of layups mentioned was demonstrated in the failure modes. The $[\pm 45/0_2/\bar{0}]_s$ plates failed horizontally through the hole in a catastrophic manner at a stress of 425 MPa (61,700 psi). One of the $[0_2/\pm 45/\bar{0}]_s$ plates, however, failed by vertical cracking along the vertical tangents to the hole boundary at 527 MPa (76,400 psi) in a noncatastrophic manner. The specimen split into two load-carrying strips which withstood an ultimate stress of 723 MPa (105,000 psi). These two characteristic failure patterns, illustrated in Fig. 3-33, point out that failure modes in some cases could be controlled by stacking-sequence variations, thus reducing the probability of catastrophic failures.

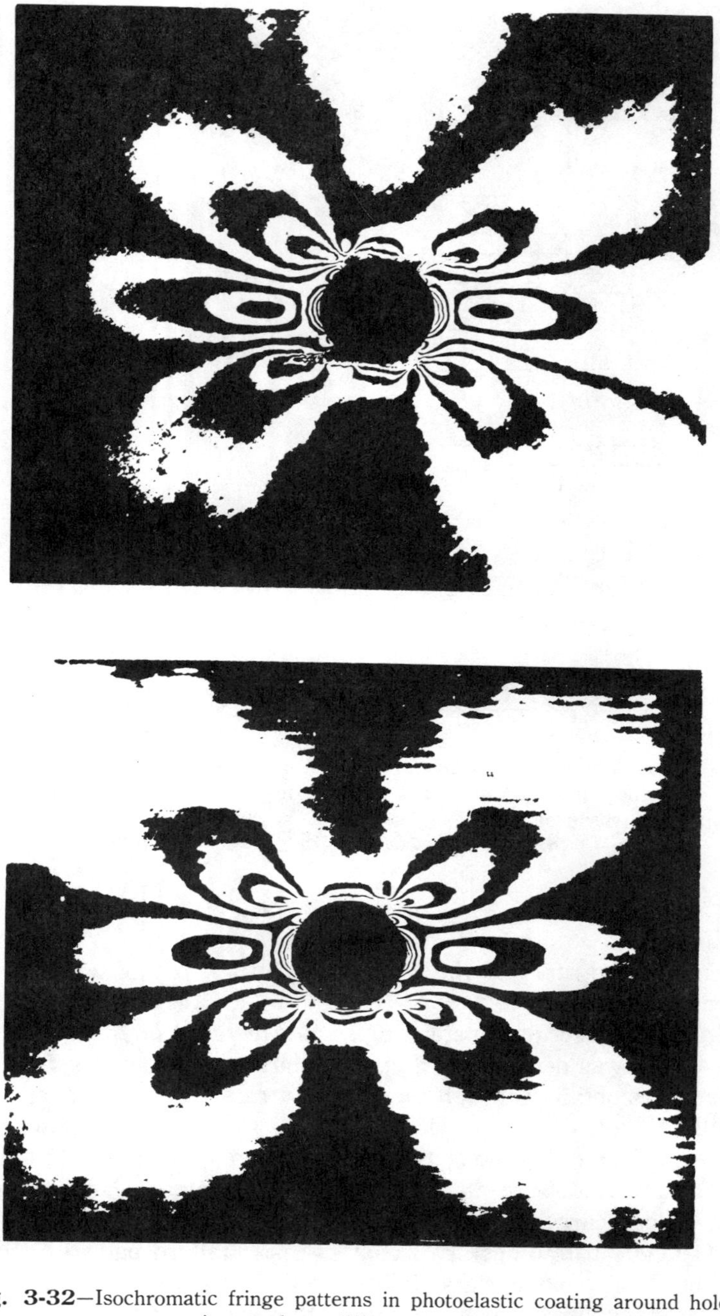

$[\pm 45/0_2/\bar{0}]_s$

$[0_2/\pm 45/\bar{0}]_s$

Fig. 3-32—Isochromatic fringe patterns in photoelastic coating around hole in boron/epoxy specimens of two different stacking sequences at an applied stress of p = 390 MPa (56,800 psi). (Ref. 5)

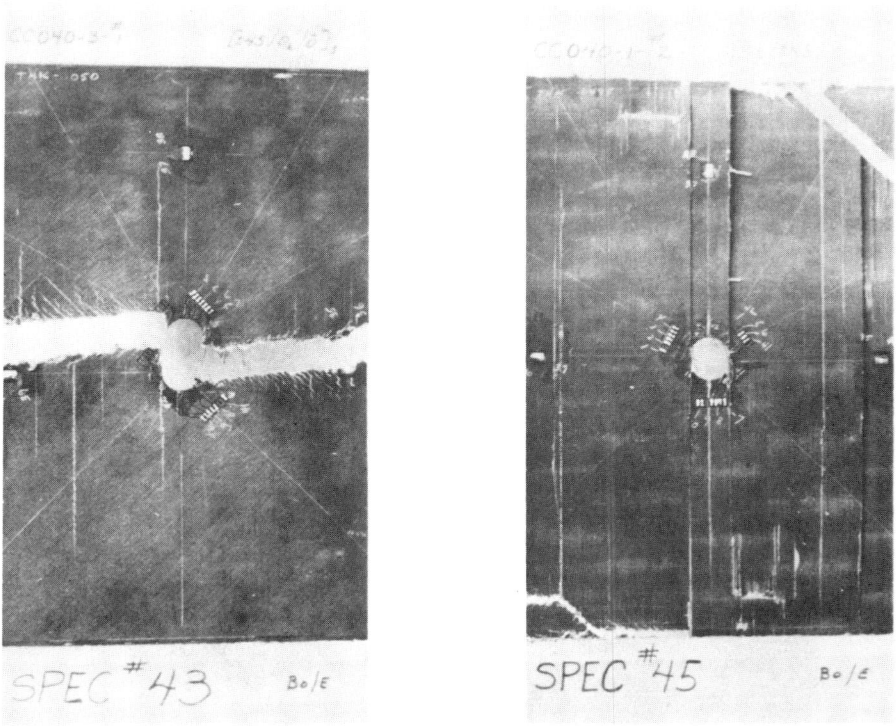

(a) (b)

Fig. 3-33—Failure patterns of boron/epoxy tensile panels with holes, (a) $[\pm 45/0_2/\bar{0}]_s$, (b) $[0_2/\pm 45/\bar{0}]_s$. (Ref. 5)

3.3.3.4 Influence of Hole Geometry

The influence of hole geometry was investigated in uniaxially loaded 25.4 cm × 66 cm (10 in. × 26 in.) boron/epoxy plates of $[0/\pm 45/0/\overline{90}]_s$ layup.[6] An elliptical hole with a 5.08 cm (2 in.) major diameter and a 2.54 cm (1 in.) minor diameter was used with the major diameter in the loading and transverse to the loading direction. A square hole of 2.54 cm × 2.54 cm (1 in. × 1 in.) dimensions with rounded corners was also tested.

Isochromatic fringe patterns at loads near faiiure are illustrated for the three hole geometries in Fig. 3-34. As in the case of the circular hole (Fig. 3-24) the concentration of fringes occurs off the horizontal axis in both elliptical shapes. In the case of the square hole the fringe concentration occurs near the corners on the vertical sides. Typical

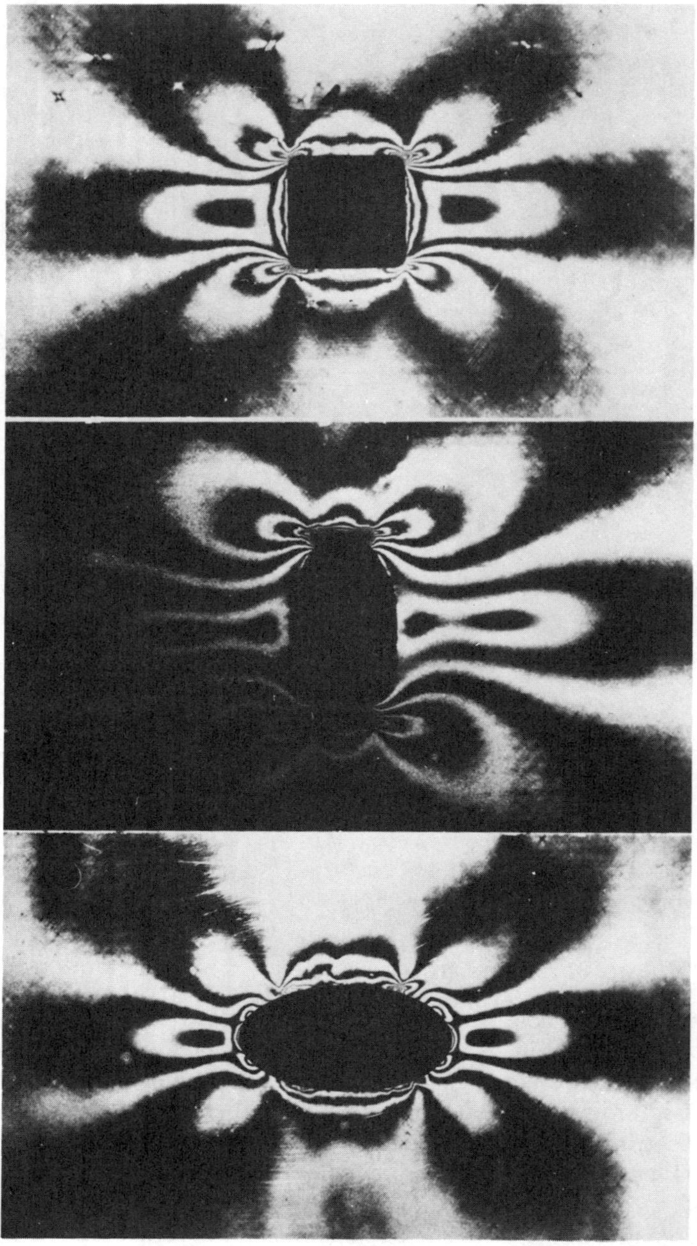

(c)

(b)

(a)

Fig. 3-34—Isochromatic fringe patterns in photoelastic coating of $[0/\pm 45/0/9\bar{0}]_s$ boron/epoxy specimens with cutouts of various shapes. Applied stress: (a) p = 361 MPa (52.3 ksi), (b) p = 241 MPa (34.9 ksi), (c) p = 331 MPa (48.0 ksi). (Ref. 6)

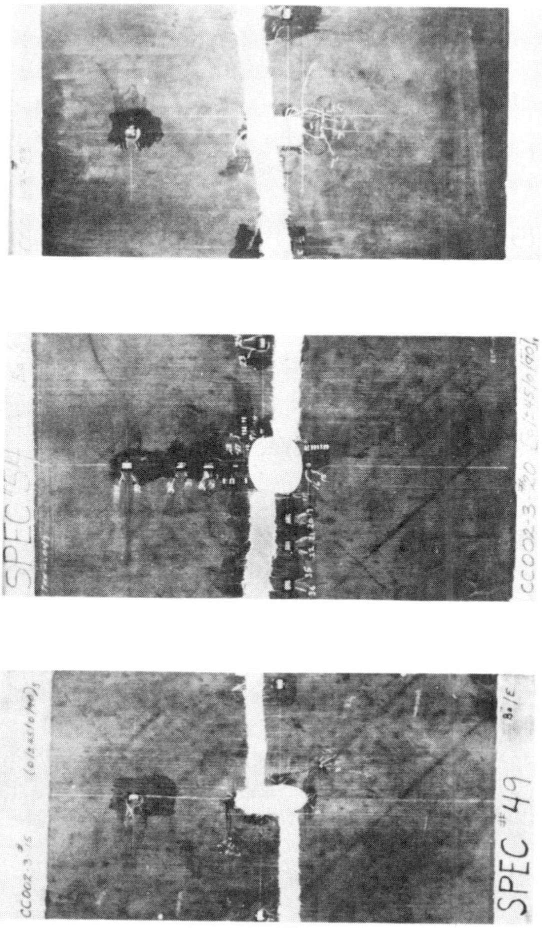

Fig. 3-35—Failure patterns of $[0/\pm 45/0/\overline{90}]_s$ boron/epoxy specimens with cutouts of various shapes. (Ref. 6)

failure patterns of these three types of specimens are shown in Fig. 3-35. Failure in all cases originates at the points of high fringe concentration. Results are tabulated in Table 3-3 where they are compared with those for the circular hole discussed before. The strength increases with decreasing stress concentration as expected. The specimen with the vertical ellipse and a stress concentration of 2.39 is approximately 27 percent stronger and the specimen with the horizontal ellipse and a stress concentration of 6.6 is approximately 17 percent weaker than the specimen with the circular hole. The panel with the square hole is stronger than those with the horizontal ellipse and the circular hole.

Guided by the photoelastic fringe patterns and the results above, it seems that the optimum cutout configuration would be a geometry incorporating some of the features of a square hole and a vertical ellipse.

3.3.3.5 Influence of Hole Diameter

The influence of hole diameter on strength of composite laminates has been discussed in Section 2.8. Two types of criteria were discussed, those based on linear elastic fracture mechanics and those based on the actual stress distributions. Photoelastic coatings can be used to elucidate both approaches. In the first approach, the value of the inherent flaw size c_o can be inferred from the area of high birefringence concentration. In the second case, photoelastic coatings help in determining stress distributions and defining the characteristic dimensions for the stress criterion.

Photoelastic coatings, along with strain gages, were used to study the deformation and failure in uniaxially loaded graphite/epoxy plates with holes and to determine the influence of hole diameter.[7,9] Isochromatic fringe patterns reveal points of high birefringence concentration on the hole boundary located between 67-deg and 71-deg from the loading axis as in the case of boron/epoxy discussed earlier. Failure, consisting of delamination and tensile cracking, was initiated at these characteristic locations. Results for specimens with holes of various diameters were in satisfactory agreement with predictions based on stress criteria. They also revealed that there is a threshold hole diameter below which the laminate is totally notch-insensitive.

3.3.3.6 Uniaxially Loaded Plates with Cracks

Deformations and failure were studied in uniaxially loaded graphite/epoxy plates with cracks and the influence of crack size on failure was determined.[7,8] The specimens were $[0/\pm45/90]_s$ laminates 12.7 cm (5 in.) wide and 56 cm (22 in.) long with transverse through-the-thickness cracks of lengths 2.54 cm (1.00 in.), 1.91 cm (0.75 in.), 1.27 cm (0.50 in.), and 0.64 cm (0.25 in.). The area around the crack was instrumented with strain gages, photoelastic coatings, and moiré grids. The application of the moiré technique in this case was discussed earlier (Section 3.2.5.4). Strain gages were applied along the horizontal (crack) axis with 0.38 mm (0.015 in.) long gages in the vicinity of the crack tip. Photoelastic coatings were 0.25 mm (0.01 in.) thick commercial coatings with a reflective backing.

The strain distribution around the crack tip and the phenomenon of damage zone formation and growth are vividly illustrated by the isochromatic fringe patterns in the photoelastic coating (Fig. 3-36). The

Table 3-3 Influence of Hole Geometry in Uniaxially Loaded $[0/\pm 45/0/\overline{90}]_s$ Boron/Epoxy Specimens

Hole Geometry	Modulus E_{xx} GPa	Modulus E_{xx} (10^6 psi)	Strain Concentration Factor, k_ϵ	Strength S_{xxT} MPa	Strength S_{xxT} (ksi)	Strength Reduction Ratio, S_{xxT}/S_o
Circular, 2.54 cm (1 in.) Diameter	115	(16.70)	3.34	291	(42.2)	0.444
Elliptical, 5.08 cm × 2.54 cm (2 in. × 1 in.), Major Axis in Direction of Loading	117	(16.97)	2.39	369	(53.5)	0.563
Elliptical, 2.54 cm × 5.08 cm (1 in. × 2 in.), Minor Axis in Direction of Loading	115	(16.67)	6.6	242	(35.0)	0.368
Square, 2.54 cm × 2.54 cm (1 in. × 1 in.)	111	(16.10)	2.9	340	(49.2)	0.518

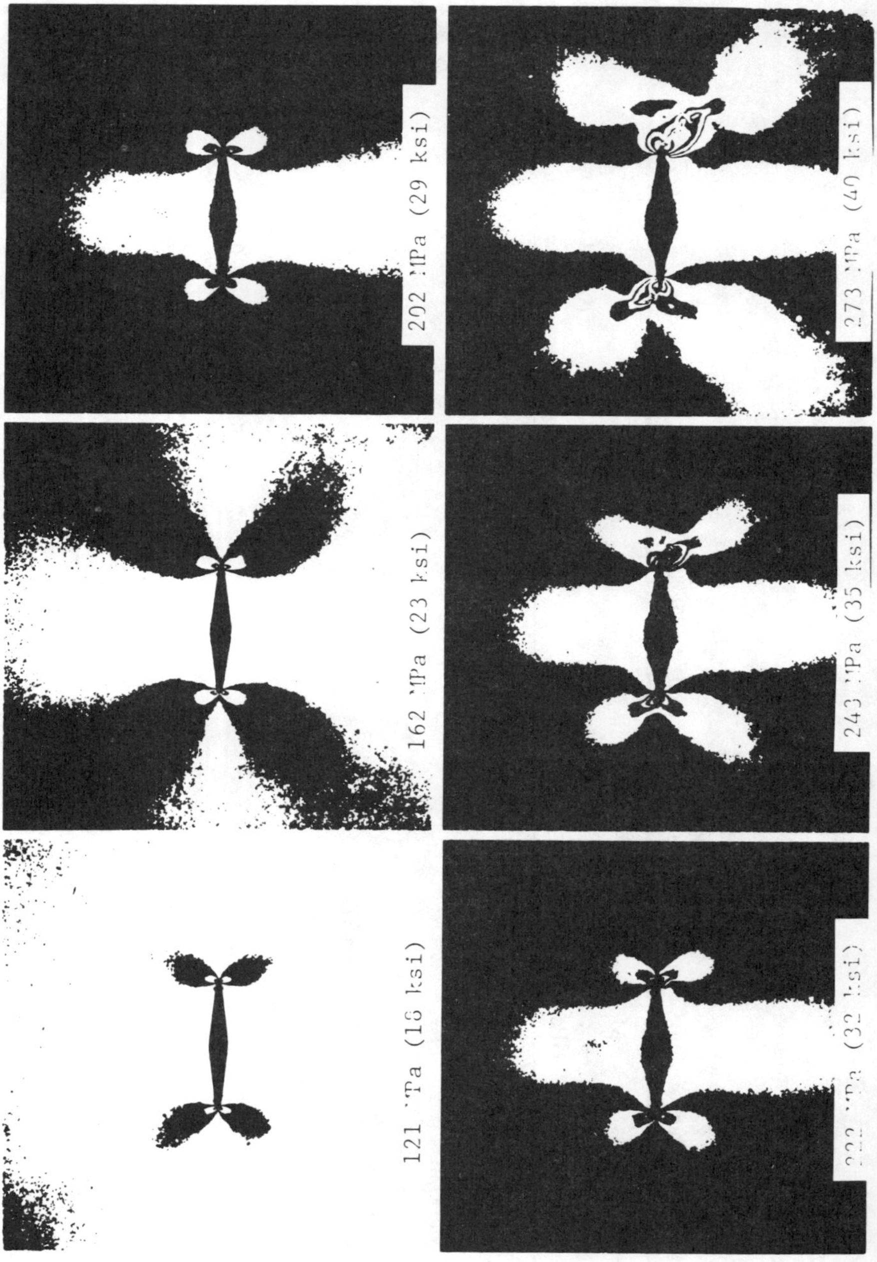

Fig. 3-36—Isochromatic fringe patterns in photoelastic coating around 1.27 cm (0.50 in.) crack of $[0/\pm 45/90]_s$ graphite/epoxy specimen at various levels of applied stress

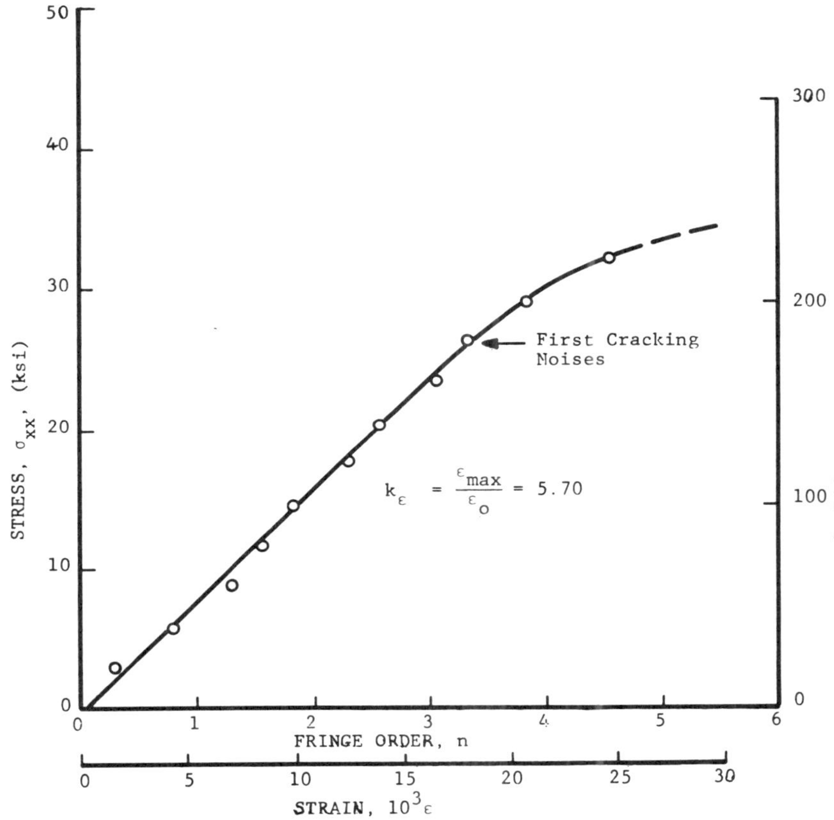

Fig. 3-37—Maximum fringe order and maximum strain at crack tip

maximum fringe order, which occurs at the crack tip off the horizontal axis, and the corresponding maximum strain are plotted in Fig. 3-37 as a function of applied stress. This curve is nearly linear up to the stress level of 180 MPa (26 ksi) when the first cracking noises were heard. The strain concentration factor computed as the ratio of the maximum strain at the crack tip to the far-field axial strain is 5.70. A noticeable characteristic is the apparent extension of the damage zone at an approximately 45-deg angle with the horizontal. As observed previously by Mandell *et al.*,[62] this damage zone consists primarily of subcracks along the fibers of individual plies, local delamination and occasional fiber breakage. The size of this zone increases with applied stress up to some critical size at which point the specimen fails catastrophically.

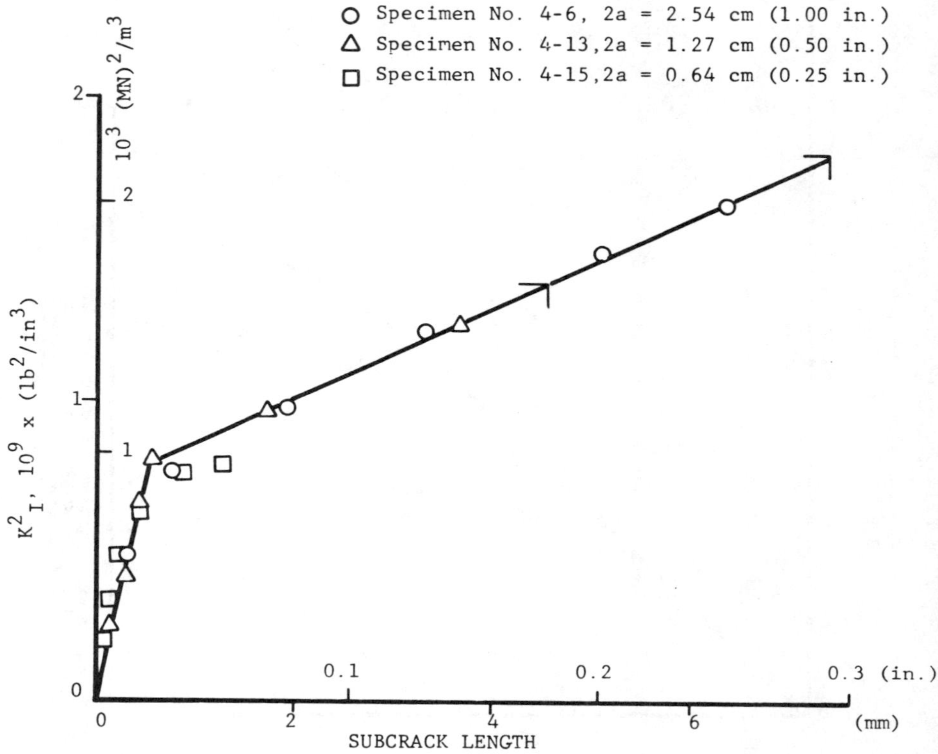

Fig. 3-38—Variation of length of subcrack with square of stress intensity factor for uniaxially loaded $[0/\pm 45/90]_s$ graphite/epoxy plates with cracks

An attempt was made to measure the damage zone and correlate it with the square of the stress intensity factor as was done by Mandell *et al.*[62] The length of the subcracks producing the damage zone was measured approximately from the photoelastic fringe pattern for three specimens with crack lengths of 2.54 cm (1.00 in.), 1.27 cm (0.50 in.) and 0.64 cm (0.25 in.). The subcrack length varies linearly with K_I^2 up to a value of $K_I \simeq 30$ MPa$\sqrt{\text{m}}$ (27.5 ksi$\sqrt{\text{in.}}$) for all three crack lengths (Fig. 3-38). Thereafter, the subcrack length increases again linearly with K_I^2 but at a faster rate which is nearly the same for all three crack lengths. This bilinear nature of the curve is characteristic of notch insensitive laminates. Failure patterns for specimens with cracks of various lengths are shown in Fig. 3-39. They all show extensive delamination near the crack tips and crack propagation that is not too straight across the width of the specimen.

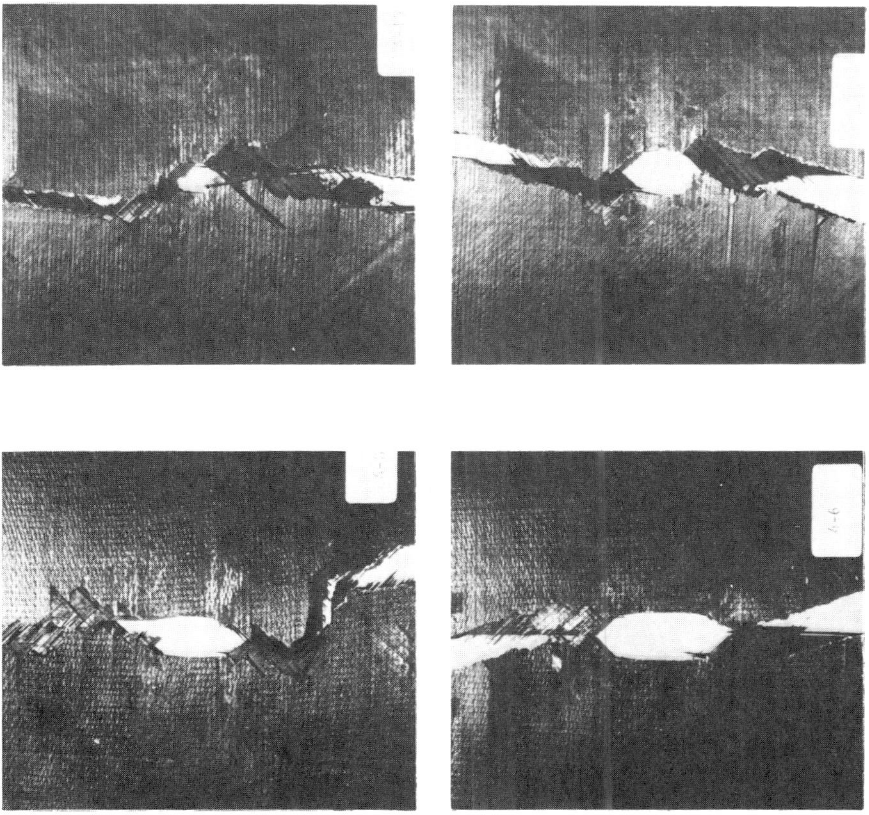

Fig. 3-39—Failure patterns in uniaxially loaded $[0/\pm 45/90]_s$ graphite/epoxy plates with cracks of various lengths [crack lengths are: 0.64 cm (0.25 in.), 1.27 cm (0.50 in.), 1.91 cm (0.75 in.) and 2.54 cm (1.00 in.)]

3.4 HOLOGRAPHIC TECHNIQUES

3.4.1 Holographic Interferometry

Holography is an optical technique based on the optical interference produced by superposition of coherent light waves reflected from the object under consideration (object beam) and those of a coherent reference beam. The laser is an ideal source of coherent monochromatic light. A typical arrangement for the holographic processes of construction and reconstruction is shown in Fig. 3-49. All the optical components shown are mounted on a vibration-isolated table. Coherent monochromatic light from a laser is divided into two beams through a beam splitter. One, the reference beam, impinges on the film plate after being reflected

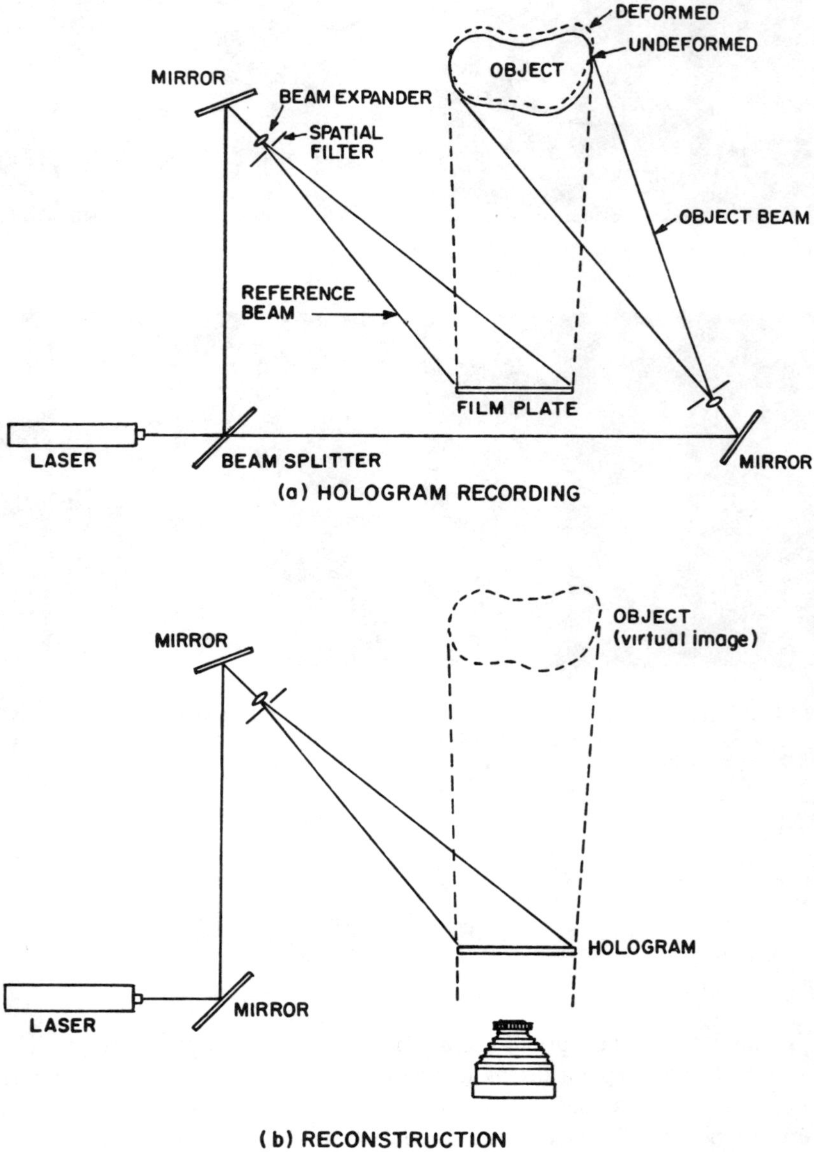

Fig. 3-40—Holographic processes

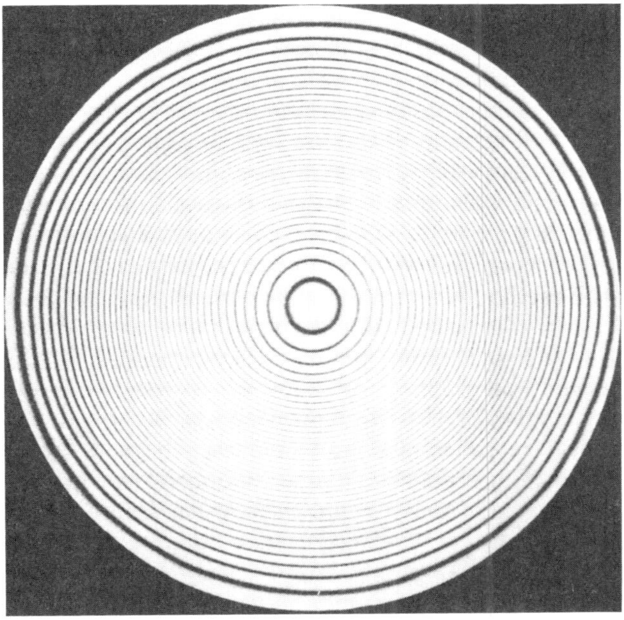

Fig. 3-41—Holographic fringe pattern of transverse deflections for a clamped circular $[0_7]$ boron/epoxy plate under uniform pressure. (Ref. 64)

by a mirror and passing through a diverging lens and a spatial filter. The other, the object beam, illuminates the object after being reflected from another mirror and passing through a diverging lens and spatial filter. This beam then scatters on the surface of the object and proceeds to the film plate where it interferes with the reference beam. The resulting interference pattern containing both amplitude and phase information of the light from the object is recorded on the film plate (hologram). In the reconstruction process, illustrated in Fig. 3-40b, the original object beam is blocked and the hologram is illuminated with the original reference beam as shown. Light refracted through the hologram contains the same intensity and phase characteristics as those of the original object. A three-dimensional virtual image of this object can be viewed or photographed through the hologram.

One of the most important applications of holography to structural and stress analysis is interferometry, i.e., the measurement of small surface displacements in a body produced by mechanical or thermal loadings. In applying holographic interferometry, two holograms of the test object are recorded on the same photographic plate with an alteration of the object surface occurring between the two recordings. The surface of the test object can be altered in several ways, including mechanical loading, pressurization, thermal loading, and acoustic vibration. Upon reconstruction, the two images will interfere with each other and will produce a set of fringes representing contours of displacement in the direction of the viewing axis. Each fringe represents a relative displacement with respect to its neighboring fringe of approximately one-half the wavelength of the light used in reconstruction. For a helium-neon laser the sensitivity of holographic interferometry is approximately 0.3 μm. In addition to the full-field nature and the high sensitivity, the method has the added advantage that it does not require any special surface preparation and can be applied to surfaces of non-optical quality such as those of composites.

An improved holographic technique which has many advantages is image-plane holography. In this approach the light waves emanating from the object are collected by an imaging lens and the film plate is placed near the image plane. This technique allows the use of white light for reconstruction and results in sharper fringe patterns. It can also be applied to larger specimens than those used in conventional holography.

Holographic techniques commonly used in experimental mechanics fall into four basic categories:

(1) Double-exposure holographic interferometry
(2) Real-time holographic interferometry
(3) Time-average or vibration holographic interferometry
(4) Dynamic double-exposure pulsed holographic interferometry.

3.4.2 Applications

Since holographic techniques are best suited for measuring out-of-plane displacements most applications to composites deal with flexure and vibration of plates. Such applications were discussed by Rowlands and Daniel,[63] Rowlands *et al.*,[64] and Maddux.[65]

3.4.2.1 Statically Loaded Plates

Double-exposure holographic interferometry has been used to measure transverse deflections and stresses in glass/epoxy, boron/epoxy, and graphite/epoxy plates. Figure 3-41 shows holographic fringe patterns representing transverse deflections in a $[0_7]$ boron/epoxy plate, 8.26 cm (3.25 in.) in diameter, clamped along its boundary, and loaded under uniform pressure.[64] The out-of-plane deflection w is given by the relation

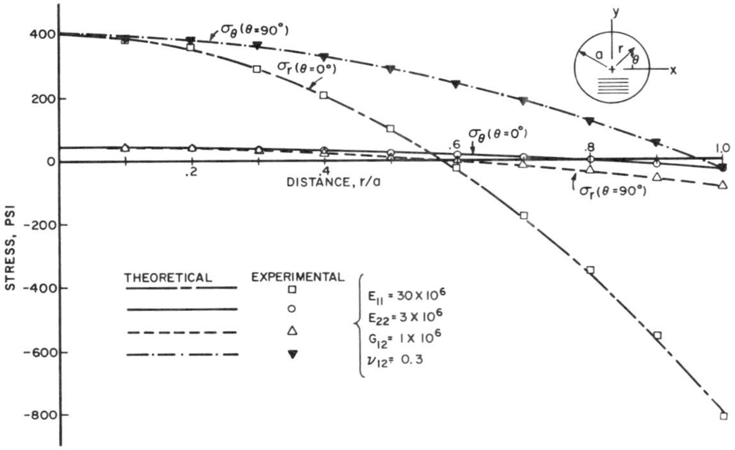

Fig. 3-42—Stresses along principal diameters in clamped circular $[0_7]$ boron/epoxy plate under uniform pressure. (Ref. 64)

$$w = \frac{2N - 1}{4}\lambda \tag{3.24}$$

where N is the holographic fringe order and λ is the wavelength of light. Displacements obtained holographically can be used to calculate inplane and interlaminar stresses in any ply of the flexed plate. In the case of the axisymmetric displacement field illustrated here, inplane stresses are determined by the following equations:

$$\begin{bmatrix} \sigma_{rr} \\ \sigma_{\theta\theta} \\ \sigma_{r\theta} \end{bmatrix} = -z \begin{bmatrix} \overline{Q}_{11} & \overline{Q}_{12} & 2\overline{Q}_{16} \\ \overline{Q}_{12} & \overline{Q}_{22} & 2\overline{Q}_{26} \\ \overline{Q}_{16} & \overline{Q}_{26} & 2\overline{Q}_{66} \end{bmatrix} \begin{Bmatrix} \dfrac{d^2w}{dr^2} \\ \dfrac{1}{r}\dfrac{dw}{dr} \\ 0 \end{Bmatrix} \tag{3.25}$$

where $[\overline{Q}]$ is the stiffness matrix referred to the polar coordinate system, and z the coordinate of the plate in the thickness direction. The derivatives of displacements needed for the stress determinations were obtained mechanically by forming moiré patterns with superimposed holographic patterns and numerically by fitting mathematical functions through discrete deflection values.[64,66] Stress distributions obtained in this manner for the unidirectional boron/epoxy plate above are shown in Fig. 3-42.[64] Results are in good agreement with theoretical predictions.

A technique for obtaining maximum and minimum strains in flexed plates directly from a hologram of the out-of-plane deflection has been described by Stetson.[67] The method consists of recording two transparent interferograms from the same reconstructed hologram of the plate. These interferograms are then rotated 180 deg with respect to each other and superimposed, producing a moiré fringe pattern at the location where identical points are superimposed. This moiré pattern has the form of conic sections (ellipses, circles or hyperbolas). The magnitude of the surface bending strains, which are proportional to the second derivatives of the out-of-plane deflections, is given by

$$\epsilon_{max} = \frac{1}{b^2} \text{ and } \epsilon_{min} = \frac{1}{a^2} \tag{3.26}$$

where a, b are the major and minor semiaxes of the conic.

3.4.2.2 Vibration Analysis

Time-average holography is ideally suited to the study of vibrating composite structures.[68] The technique provides a contour map of vibration amplitudes of a structure vibrating at one of its resonant frequencies. The resonant frequencies are found by using real-time holography. First, a hologram is made of the object at rest, processed and kept in the same place. Then, the object is excited with a speaker or a piezoelectric crystal. As the frequency is varied the interference fringe patterns corresponding to vibration modes and amplitudes can be viewed in real time. After the resonant frequencies have been identified the experiment is repeated and time-average holograms are made at each resonant frequency.

With time-average holography applied to a sinusoidally vibrating object with the exposure duration long compared to the period of vibration, the amplitude of vibration $w^o(x,y)$ is related to the roots of the zero-order Bessel function J_o. That is, a dark fringe is formed when the following relation is satisfied

$$J_o(4\pi w^o/\lambda) = 0 \tag{3.27}$$

where

$$w(x,y,t) = w^o(x,y) \sin \omega t \tag{3.28}$$

The incident and reflected beams are assumed normal to the surface of the object.

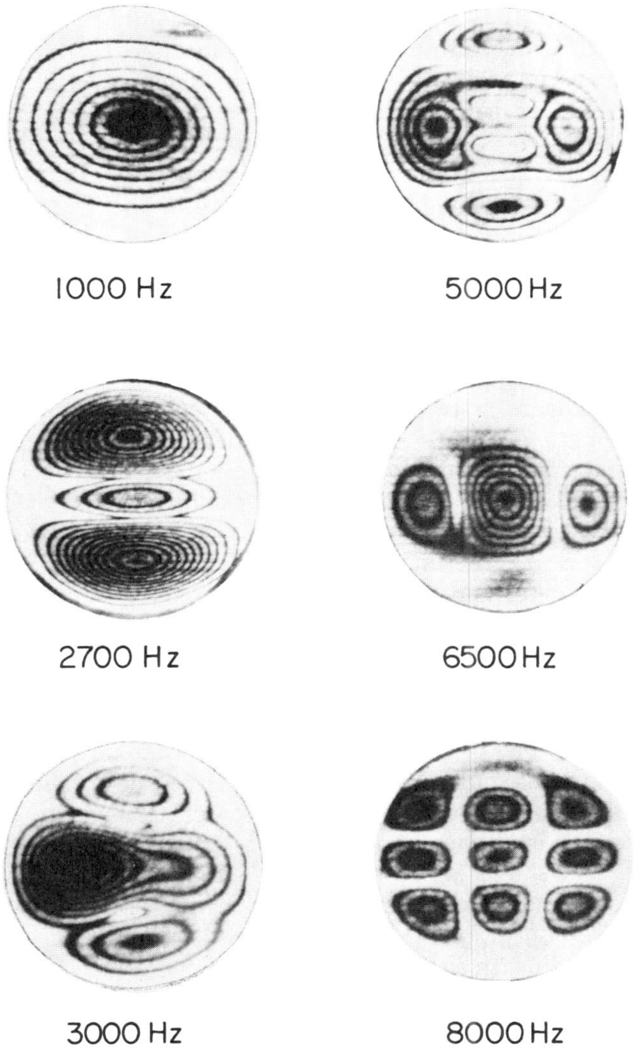

Fig. 3-43 —Vibrational shapes of a $[0_7]$ unidirectional boron/epoxy plate. (Ref. 63)

Applications of vibration holography to isotropic structures have been described amply in the literature.[65,68,69] Applications to composites in principle are the same, but the resonant modes are in general different. Figure 3-43 shows several vibrational modes and combinations of modes for a circular unidirectional $[0_7]$ boron/epoxy plate.[63] The plate, 8.25 cm

(3.25 in.) in diameter was clamped on the boundary and excited at the center. The anisotropy of the material alters significantly the axisymmetric response associated with an isotropic plate. The first mode at 1000 Hz shows elliptical contours with the major axes parallel to the fiber direction. The pattern is similar to the deflection pattern obtained for concentrated static loading at the center. The holographic patterns can be analyzed quantitatively to obtain amplitudes of vibration and deflection distribution, and, by differentiation of the deflections, stress distributions.[64]

3.4.2.3 Nondestructive Evaluation

An important application of holography is the nondestructive evaluation (NDE) of material integrity. The technique relies on the fact that if the state of stress in the component is changed, the surface displacements will be altered locally around surface or near-surface defects. This alteration is manifested as a local anomaly in the overall fringe pattern. This nondestructive detection is carried out either by double-exposure or real-time holography. Cracking and delamination in composite laminates and bond defects in composite sandwich panels or adhesive joints can be detected. The type of loading applied to reveal the presence of a defect depends on the type of defect sought, the material properties and the component geometry.

Disbonds in honeycomb sandwich panels can be detected by thermal stressing, vacuum or pressure loading, and vibrational excitation of the panel.[65,69–73] Double-pulse holography was also used to detect disbonds and delaminations in composite compressor blades, by subjecting them to acoustic excitation.[74]

Figure 3-44 shows an example of the application of holography to the detection of fatigue-induced damage growing around a circular hole in a $[(0/\pm45/90)_s]_2$ graphite/epoxy laminate.[75] The specimen was 2.54 cm (1.00 in.) wide with a 0.48 cm (0.19 in.) diameter central hole. It was loaded for 118,267 cycles to a maximum load of 12,500 N (2810 lb) and then inspected. Holographic fringe patterns were induced thermally by using a radiant heat source to raise the surface temperature of the specimens by up to 10°C above ambient room temperature. The optimum thermal loading was obtained by real-time holography but the fringe patterns of Fig. 3-44 were obtained by double-exposure holography. Figure 3-44a shows the thermally induced fringe patterns on the front surface of the specimen, while Fig. 3-44b shows similar patterns on the back surface. These fringe patterns help locate delaminations and some surface-ply matrix cracks. Some of the defects detected are: (a) matrix cracks on the front and back surface 0-deg

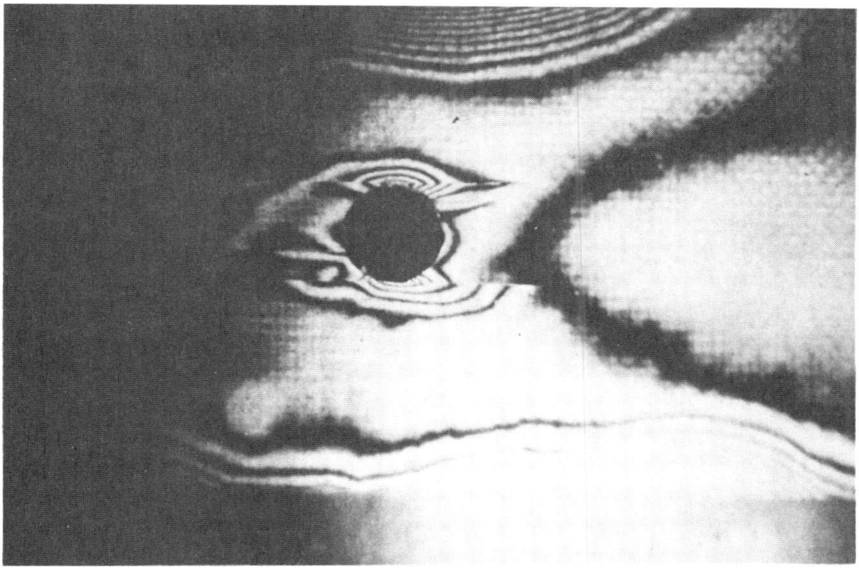

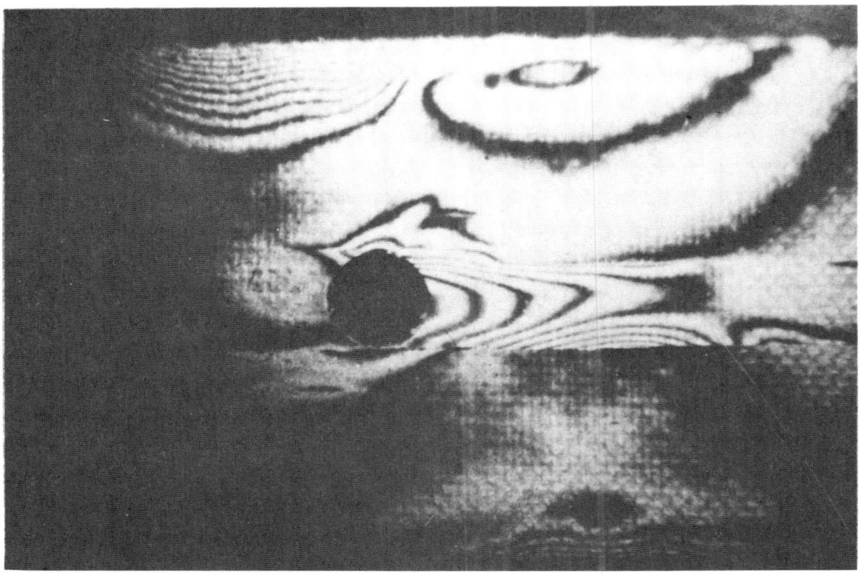

Fig. 3-44 —Thermally induced holographic fringe patterns in fatigue loaded $[(0/\pm 45/90)_s]_2$ graphite/epoxy specimen with circular hole. (a) front surface, (b) back surface (as viewed from the front through the specimen). (Ref. 75)

plies detected by the fringe pattern anomalies along vertical lines, (b) delaminations between front surface 0-deg ply and adjacent 45-deg ply on the sides and below the hole, (c) delaminations between the back surface 0-deg ply and adjacent 45-deg ply on the sides and above the hole, and (d) delaminations between the 90-deg and adjacent 45-deg plies near the edges.

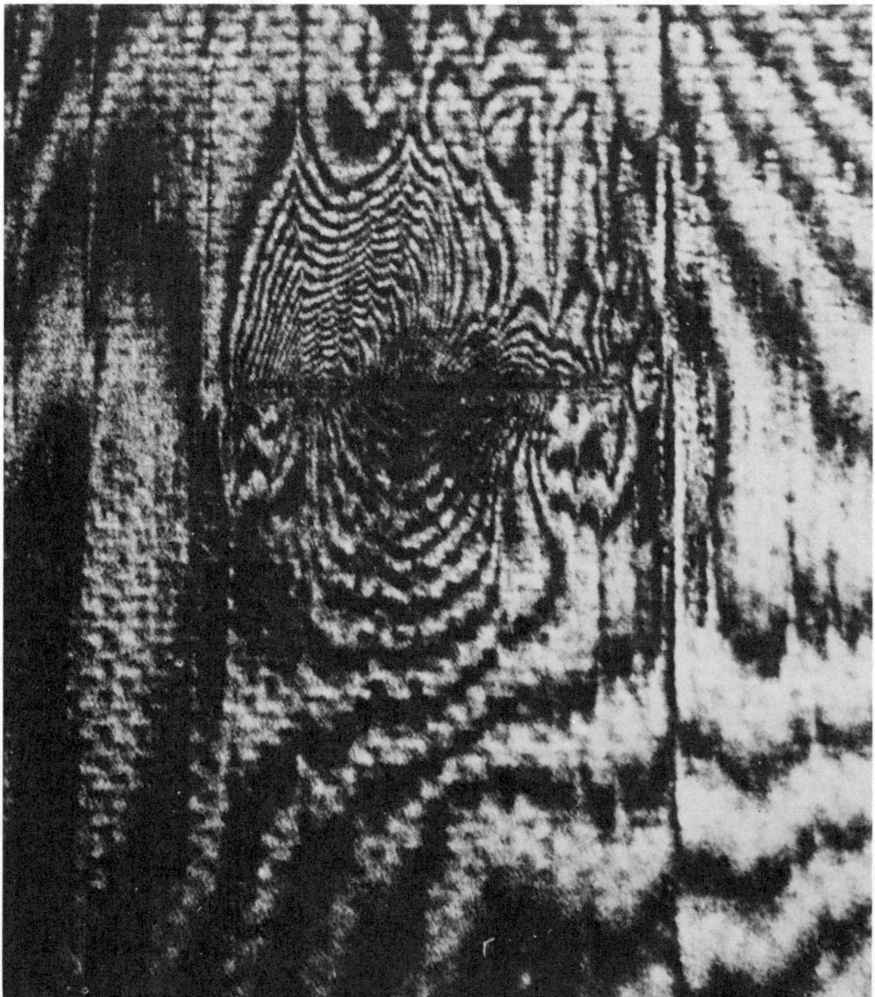

Fig. 3-45—Holographic fringe patterns in statically loaded $(0^K/\pm45^C/0^K)_s$ Kevlar/graphite/epoxy hybrid specimen with cracks. (Ref. 75)

Another example of damage detection around a crack is illustrated in Fig. 3-45.[75] The specimen was a $(0^K/\pm45^C/0^K)_s$ Kevlar/graphite/epoxy hybrid with the 0-deg and ±45-deg plies being Kevlar/epoxy and graphite/epoxy, respectively. It was 2.54 cm (1.00 in.) wide, 1.14 mm (0.045 in.) thick and had a 1.05 cm (0.413 in.) long slit perpendicular to the loading (0-deg) direction. It also had a manufacturing defect (gap between the fibers on back surface). It was loaded statically to a gross section stress of 312 MPa (45.2 ksi), unloaded and then inspected holographically. The damage in the neighborhood of the slit is extensive, consisting of matrix cracks, delaminations and fiber breaks. The matrix cracks on the surface plies appear as cusps in the holographic fringes, delaminations appear as anomalous and dense fringe patterns and the manufacturing defect appears as a vertical line in the fringe pattern.

Holography is an effective NDE technique for composite materials, capable of detecting delaminations and cracks in the surface plies.[76] It is capable of detecting both matrix cracks and fiber breaks in the surface plies and can give some information relative to the through-the-thickness distribution of delaminations. It does not, however, detect subsurface matrix cracking. It is capable of detecting flaws near free edges but not subsurface ones away from the edges. The potential of the method has not been fully developed yet. More work is necessary to interpret fringe patterns produced by a variety of excitation techniques (flaws).

3.4.2.4 Material Properties

Real-time holographic techniques were used by Freund to measure micromechanical properties of quasi-isotropic graphite/epoxy laminates and honeycomb sandwich panels with graphite/epoxy face plates.[77] Yield and creep characteristics at low stress levels (microyield and microcreep) were studied by centrally loading a circular plate of the material supported at three equally spaced points on the periphery. The stress level at which 0.025 μm (1 μin.) of the permanent deformation was observed (microyield) was established.

3.5 SPECKLE INTERFEROMETRIC TECHNIQUES

Speckle interferometry makes use of the speckle pattern produced on the surface of an object illuminated by coherent light. It has many characteristics complementary to those of holographic interferometry. It is less sensitive than the latter and is primarily indicative of inplane displacements.

Leendertz described a technique in which the object surface is illuminated by two coherent beams equally inclined from the normal to the surface.[78] An alternative method has been proposed by Archbold *et al.*, in which the object is illuminated by one beam and two exposures are recorded, before and after loading.[79] The techniques mentioned were further generalized by Hung and Hovanesian to three-dimensional cases.[80]

The most important development in this area is that of speckle-shearing interferometry introduced by Hung and Taylor.[81] The technique allows direct determination of derivatives of surface displacements. It overcomes or alleviates many of the limitations of conventional interferometry, namely: (1) the setup is relatively simple and it does not require laborious alignment of optics, (2) mechanical and environmental stability is not critical, (3) coherent length of light is minimized, (4) sensitivity can be controlled over a wider range, (5) films of much lower resolution can be used, (6) the fringes always localize on the specimen surface, and (7) strains can be recorded directly without the need of differentiating displacements. More recent developments introduced by Hung *et al.*, allow the simultaneous determination of derivatives of surface displacements along any direction and with variable sensitivity using a single photographic record (shearing specklegram).[82,83]

Speckle-shearing interferometry can be applied effectively to testing of composite materials. Slopes of deflection can be obtained directly in studies of flexural deformations of beams, plates and shells. Inplane strains can be measured over the whole field in a manner equivalent to a full-field strain gage. Applications to vibration problems are similar to those of holographic interferometry. The time-integrated fringes obtained, however, depict slopes of modal amplitude of vibration. The feasibility of the method as a nondestructive evaluation tool has been demonstrated, but the full potential of the method has not been fully explored yet.[84]

3.6 ANISOTROPIC PHOTOELASTICITY—BIREFRINGENT COMPOSITES

3.6.1 Stress Optic Law

Several investigators have conducted both analytical and experimental studies in an effort to develop stress-optic relations for birefringent fiber-reinforced composite materials. Pih and Knight[85] pioneered work in this area and developed a stress-optic law based upon a stress-proportioning technique. Later, Sampson[86] formulated a stress-optic law which hypothesized the concept of Mohr's circle of birefringence. In addition, Sampson introduced the concept that three photoelastic constants are required in order to characterize these new materials photoelastically. Dally and Prabhakaran[87] developed simple methods for the fabrication of the fiber-reinforced birefringent material, and predicted the three fundamental photoelastic constants based upon properties of the constituents. Both experimental and theoretical results presented closely agreed with the stress-optic law formulated by Sampson. Bert[88] has shown that the concept of a Mohr's circle of birefringence, as proposed by Sampson, is a direct result of the tensorial nature of birefringence. Pipes and Rose expressed a strain-optic law based on one strain fringe value alone.[89] Prabhakaran examined and verified Sampson's stress-optic law under biaxial stress conditions.[90] Cernosek recently showed the equivalence of Sampson's phenomenological theory and the stress-proportioning concept.[91] He also showed that the optical isoclinic parameter can be predicted accurately, even in the presence of residual birefringence. More recently, Knight and Pih formulated general stress- and strain-optic laws by using tensor forms of stress, birefringence and stress fringe values.[92] These relations were then simplified to the more familiar two-dimensional equations.

The stress-optic law proposed by Sampson is developed as follows:[86]

The birefringence per unit thickness in an isotropic material is given by

$$N = \frac{p - q}{f} \tag{3.29}$$

where p, q = major and minor principal stresses

f = fringe value

The relation above, when referred to an arbitrary x-y system of coordinates, takes the form

$$N = \sqrt{\left(\frac{\sigma_x}{f} - \frac{\sigma_y}{f}\right)^2 + \left(\frac{2\tau_{xy}}{f}\right)^2} \tag{3.30}$$

Based on this form, Sampson postulated the following law for an orthotropic material:

$$N = \sqrt{\left(\frac{\sigma_x}{f_x} - \frac{\sigma_y}{f_y}\right)^2 + \left(\frac{2\tau_{xy}}{f_{xy}}\right)^2} \tag{3.31}$$

which involves three different material fringe values, f_x, f_y and f_{xy}. If the x-y system is assumed to coincide with the principal material axes of the orthotropic material, the components of birefringence

$$N_1 = \frac{\sigma_1}{f_1}$$

$$N_2 = \frac{\sigma_2}{f_2} \qquad (3.32)$$

$$N_{12} = \frac{\tau_{12}}{f_{12}}$$

follow the same transformation relations as the corresponding stress components.

From considerations of equilibrium and stress-transformation relations, eq (3.31) yields

$$f_\theta = f_1 \left[\left(\cos^2\theta - \frac{f_1}{f_2}\sin^2\theta\right)^2 + \frac{f_1^2}{f_{12}^2}\sin^2 2\theta\right]^{-1/2} \qquad (3.33)$$

where θ is the angle between the principal stress p and the 1-direction. By determining f_1, f_2 and f_{45} eq (3.33) can be solved for f_{12} by substituting $\theta = 45$ deg.

In introducing the tensorial nature of birefringence Bert[88] proposed that the general unidirectional fiber-reinforced composite material could be considered similar to an orthorhombic crystal which possesses three principal material directions. Thus, the components of the birefringence tensor N_i are related to the stress components in the lamina coordinate system as follows:

$$N_1 = B_{11}\sigma_1 + B_{12}\sigma_2 + 0\sigma_6$$

$$N_2 = B_{21}\sigma_1 + B_{22}\sigma_2 + 0\sigma_6 \qquad (3.34)$$

$$N_6 = 0\sigma_1 + 0\sigma_2 + B_{66}\sigma_6$$

where the B_{ij} are photoelastic constants.

The familiar stress-optic coefficients can be expressed in terms of the photoelastic constants, B_{ij}.

$$f_1 = (B_{11} - B_{21})^{-1}$$

$$f_2 = (B_{22} - B_{12})^{-1} \qquad (3.35)$$

$$f_{12} = B_{66}^{-1}$$

The isoclinic angle is given by

$$\tan 2\phi = \frac{-2B_{66}\tau_{12}}{(B_{11} - B_{21})\sigma_1 - (B_{22} - B_{12})\sigma_2} = \frac{-2\tau_{12}/f_{12}}{\sigma_1/f_1 - \sigma_2/f_2} \tag{3.36}$$

3.6.2 *Strain-Optic Law*

The strain-optic law may be derived from the stress-optic law through substitution of the constitutive relations into eq (3.31).

The strain optic coefficients f_1^ϵ, f_2^ϵ, and f_{12}^ϵ are given in terms of the stress optic coefficients and the material stiffness constants Q_{ij}.

$$\begin{aligned} f_1^\epsilon &= \frac{f_1 f_2}{f_2 Q_{11} - f_1 Q_{12}} \\ f_2^\epsilon &= \frac{f_1 f_2}{f_1 Q_{22} - f_2 Q_{12}} \\ f_{12}^\epsilon &= \frac{f_{12}}{2Q_{66}} \end{aligned} \tag{3.37}$$

The isoclinic angle may therefore be related to the strain components and strain optic coefficients as follows:

$$\tan 2\phi = \frac{\gamma_{12}/f_{12}}{[\epsilon_1/f_1^\epsilon - \epsilon_2/f_2^\epsilon]} \tag{3.38}$$

A simplified strain-optic law was developed by Pipes and Rose[89] in which the birefringence response of a photoelastic composite is assumed to obey the classical isotropic strain-optic law.

$$\epsilon_p - \epsilon_q = \frac{nf_\epsilon}{t} = Nf_\epsilon \tag{3.39}$$

where ϵ_p, ϵ_q are the principal strains; n is the isochromatic-fringe order; f_ϵ is the strain-optic coefficient; and t is the material thickness.

Using strain transformation and stress–strain relations we obtain the following strain-optic law

$$N^2 = [\frac{(S_{11} - S_{12})}{f_\epsilon}\sigma_x - \frac{(S_{22} - S_{12})}{f_\epsilon}\sigma_y]^2 + [\frac{S_{66}\tau_{xy}}{f_\epsilon}]^2 \tag{3.40}$$

where S_{ij} are the material compliance constants. By comparing this result with Sampson's stress-optic law [eq (3.31)] we obtain the following relationships between the three fundamental fringe values and the single strain fringe value:

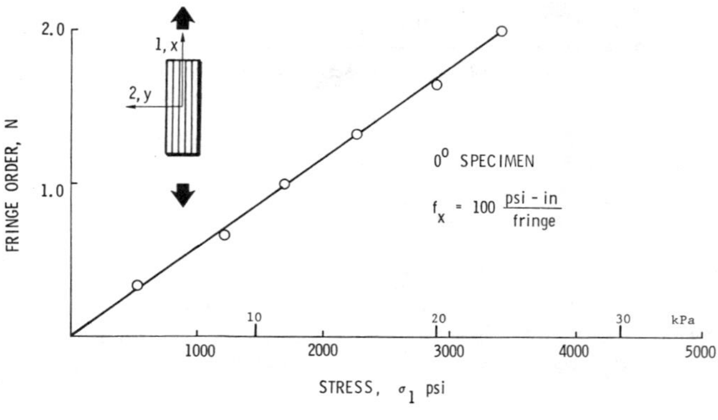

Fig. 3-46—Optical characterization, 0-deg test

$$
\begin{aligned}
f_1 &= \frac{f_\epsilon}{(S_{11} - S_{12})} \\
f_2 &= \frac{f_\epsilon}{(S_{22} - S_{12})} \qquad (3.41) \\
f_{12} &= \frac{2f_\epsilon}{S_{66}}
\end{aligned}
$$

The relationship between the S_{ij} constants and engineering properties and their determination are discussed in Sections 2.4.2 and 4.3.1.

3.6.3 Optical and Mechanical Characterization

An orthotropic birefringent material can be fully characterized optically by determining the fringe values f_1, f_2, and f_{12}. The first two can be easily determined by testing tensile coupons with the load at 0 deg and 90 deg with respect to the fiber direction. The value of f_{12} is determined indirectly by testing a tensile coupon at 45 deg to the fiber direction and using eq (3.33).

Birefringent properties were obtained for an *S*-glass/epoxy composite with anisotropy ratios of $E_1/E_2 = 2.1$ and $G_{12}/E_2 = 0.32$.[89] Stress-birefringence curves for the three basic tests are shown in Figs. 3-46, 47, and 48. The stress fringe values obtained from these tests are:

f_1 = 17.5 MPa-m/fringe (100 psi-in./fringe)
f_2 = 10.5 MPa-m/fringe (60 psi-in./fringe)
f_{45} = 7.0 MPa-m/fringe (40 psi-in./fringe)
f_{12} = 6.9 MPa-m/fringe (39.6 psi-in./fringe)

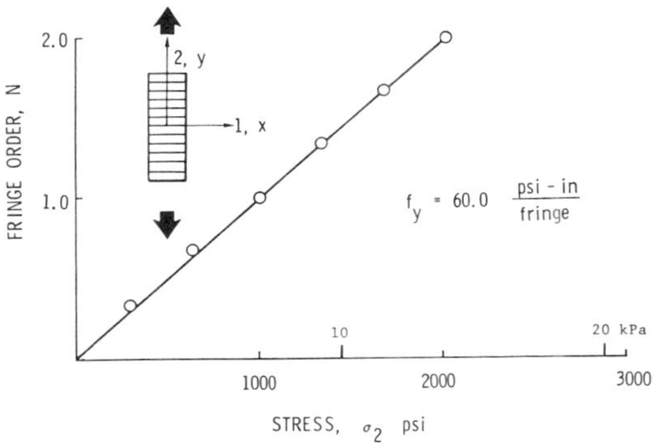

Fig. 3-47—Optical characterization, 90-deg test

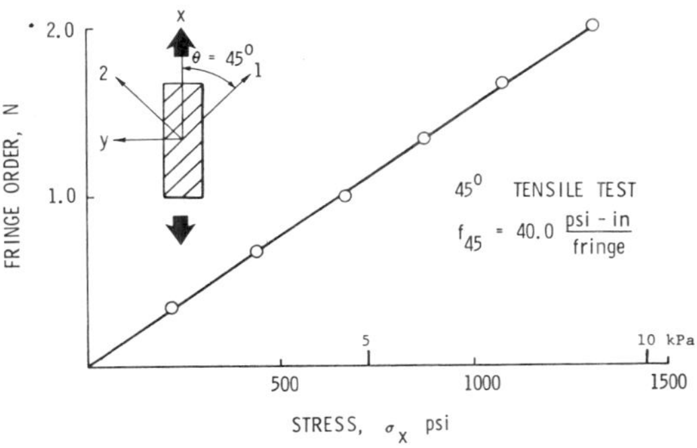

Fig. 3-48—Optical characterization, 45-deg test

The corresponding mechanical properties obtained by similar coupon tests are:

E_1 = 2.622 GPa (0.38 × 16^6 psi)
E_2 = 1.346 GPa (0.195 × 10^6 psi)
G_{12} = 0.386 GPa (0.056 × 10^6 psi)
ν_{12} = 0.33

The compliance and stiffness terms S_{ij} and Q_{ij} were computed from these engineering properties. Substituting the values for the stiffness terms and the fringe values obtained from the characterization tests in eq (3.41) we obtain:

f_1^ϵ = 8.76 μm/fringe (345 μin./fringe)
f_2^ϵ = 9.17 μm/fringe (361 μin./fringe)
f_{12}^ϵ = 8.74 μm/fringe (344 μin./fringe)

These results show that the strain fringe values differ by less than five percent for this material system. Therefore, it follows from eq (3.38) that the isoclinic angle nearly coincides with the principal strain direction.

Prabhakaran[93] pointed out that for the material he studied, the variation of stress optic coefficients and elastic constants with fiber volume fraction are quite similar. Therefore, for such materials the simplified strain-optic law could provide a reasonable approximation over a wide range in fiber volume fraction and hence material anisotropy.

REFERENCES

1. Dally, J.W. and Riley, W.F., *Experimental Stress Analysis,* McGraw-Hill, New York, 366-390 (1965).
2. Rowlands, R.E., Daniel, I.M. and Whiteside, J.B., "Stress and Failure Analysis of a Glass-Epoxy Plate with a Hole," *Experimental Mechanics,* Vol. 13, No. 1, 31-37 (Jan. 1973).
3. Daniel, I.M., Rowlands, R.E. and Whiteside, J.B., "Deformation and Failure of Boron-Epoxy Plate with Circular Hole," *Analysis of the Test Methods for High Modulus Fibers and Composites,* ASTM STP 521, American Society for Testing and Materials, 143-164 (1973).
4. Tirosh, J. and Berg, C.A., "Experimental Stress Intensity Factor Measurements in Orthotropic Composites," *Composite Materials: Testing and Design* (Third Conference), ASTM STP 546, American Society for Testing and Materials, 663-673 (1974).
5. Daniel, I.M., Rowlands, R.E. and Whiteside, J.B., "Effects of Material and Stacking Sequence on Behavior of Composite Plates with Holes," *Experimental Mechanics,* Vol. 14, No. 1, 1-9 (Jan. 1974).
6. Rowlands, R.E., Daniel, I.M. and Whiteside, J.B., "Geometric and Loading Effects on Strength of Composite Plates with Cutouts," *Composite Materials; Testing and Design* (Third Conference), ASTM STP 546, American Society for Testing and Materials, 361-375 (1974).
7. Daniel, I.M., "Biaxial Testing of Graphite/Epoxy Composites Containing Stress Concentrations," AFML-TR-76-244, Part I (Dec. 1976); Part II (June 1977).
8. Daniel, I.M., "Strain and Failure Analysis in Graphite/Epoxy Plates with Cracks," *Experimental Mechanics,* Vol. 18, No. 7, 246-252 (July 1978).
9. Daniel, I.M., "The Behavior of Uniaxially Loaded Graphite/Epoxy Plates with Holes," *Proceedings of Second International Conf. on Composite Materials,* ICCM/2, Toronto, Canada, 1019-1034 (April 1978).
10. Daniel, I.M., "Deformation and Failure of Composite Laminates with Cracks in Biaxial Stress Fields," *Proc. of Sixth International Conference on Experimental Stress Analysis,* VDI-Berichte Nr. 313, Munich, West Germany, 705-710 (Sept. 1978).
11. Daniel, I.M., "Behavior of Graphite/Epoxy Plates with Holes Under Biaxial Loading," *Experimental Mechanics,* Vol. 20, No. 1, 1-8 (Jan. 1980).

12. Daniel, I.M., Mullineaux, J.L., Ahimaz, F.J. and Liber, T., "The Embedded Strain Gage Technique for Testing Boron/Epoxy Composites," *Composite Materials: Testing and Design* (Second Conference), ASTM STP 497, American Society for Testing and Materials, 257-272 (1972).
13. Daniel, I.M., Liber, T. and Chamis, C.C., "Measurement of Residual Strains in Boron/Epoxy and Glass/Epoxy Laminates," *Composite Reliability,* ASTM STP 580, American Society for Testing and Materials, 340-351 (1975).
14. Daniel, I.M. and Liber, T., "Lamination Residual Stresses in Fiber Composites," IITRI Report D6073-I, NASA CR-134826 (March 1975).
15. Daniel, I.M. and Liber, T., "Lamination Residual Stresses in Hybrid Composites," IITRI Report D6073-II, NASA CR-135085 (June 1976).
16. Daniel, I.M. and Liber, T., "Measurement of Lamination Residual Strains in Graphite Fiber Laminates," *Proc. of Second International Conference on Mechanical Behavior of Materials,* ICM-II, Boston, MA, 214-218 (Aug. 16-20, 1976).
17. Daniel, I.M. and Liber, T., "Effect of Laminate Construction on Residual Stresses in Graphite/Polyimide Composites," *Experimental Mechanics,* Vol. 17, No. 1, 21-25 (Jan. 1977).
18. Daniel, I.M. and Liber, T., "Lamination Residual Stresses in Hybrid Laminates," *Composite Materials: Testing and Design* (Fourth Conference), ASTM STP 617, American Society for Testing and Materials, 331-343 (1977).
19. Daniel, I.M. and Liber, T., "Wave Propagation in Fiber Composite Laminates," IITRI Report D6073-III, NASA CR-135086 (June 1976).
20. Daniel, I.M. and Liber, T., "Testing of Composites at High Strain Rates," *Proc. of Second International Conference on Composite Materials,* ICCM/2, Toronto, Canada, 1003-1018 (April 1978).
21. Daniel, I.M., Liber, T. and LaBedz, R., "Wave Propagation in Transversely Impacted Composite Laminates," *Experimental Mechanics,* Vol. 19, No. 1, 9-16 (Jan. 1979).
22. Daniel, I.M., LaBedz, R. and Liber, T., "New Method for Testing Composites at Very High Strain Rates," *Experimental Mechanics,* Vol. 21, No. 2, 71-77 (Feb. 1981).
23. Freeman, W. and Campbell, M.D., "Thermal Expansion Characteristics of Graphite Reinforced Composite Materials," *Composite Materials: Testing and Design* (Second Conference), ASTM STP 497, American Society for Testing and Materials, 121-142 (1972).
24. Wang, A.S.D., Pipes, R.B. and Ahmadi, A., "Thermoelastic Expansion of Graphite-Epoxy Unidirectional and Angle-Ply Composites," *Composite Reliability,* ASTM STP 580, American Society for Testing and Materials, 574-585 (1975).
25. Daniel, I.M., "Thermal Deformations and Residual Stresses in Fiber Composites," *Proc. of 1977 International Symposium on Thermal Expansion of Solids,* Hecla Island, Manitoba, Canada (Aug. 29-31, 1977); *Thermal Expansion 6,* ed. by Ian D. Peggs, Plenum Publishing Corp., New York, 203-221 (1978).
26. Chamis, C.C., "Lamination Residual Stresses in Cross-Plied Fiber Com-

posites," *Proc. of 26th Annual Conference of SPI,* Reinforced Plastics/Composites Division, Paper No. 17-D (Feb. 1971).

27. Hahn, H.T. and Pagano, N.J., "Curing Stresses in Composite Laminates," *J. Composite Materials,* Vol. 9, 91-106 (1975).
28. Hahn, H.T., "Residual Stresses in Polymer Matrix Composite Laminates," *J. Composite Materials,* Vol. 10, 266-278 (Oct. 1976).
29. Daniel, I.M. and Liber, T., "Relaxation of Residual Stresses in Angle-Ply Composite Laminates," *Composite Materials: The Influence of Mechanics of Failure on Design,* Army Symposium on Solid Mechanics, 1976, South Yarmouth, MA (Sept. 1976).
30. Sciammarella, C.A., "Techniques of Fringe Interpolation in Moiré Patterns," *Experimental Mechanics,* Vol. 7, No. 11, 19A-30A (1967).
31. Sciammarella, C.A. and Sturgeon, D.L., "Digital-Filtering Techniques Applied to the Interpolation of Moiré-Fringe Data," *Experimental Mechanics,* Vol. 7, No. 11, 468-475 (1967).
32. Sampson, R.C. and Campbell, D.M., "The Grid-Shift Technique for Moiré Analysis of Strain in Solid Propellants," *Experimental Mechanics,* Vol. 7, No. 11, 449-457 (Nov. 1967).
33. Chiang, F.P., Parks, V.J. and Durelli, A.J., "Moiré-Fringe Interpolation and Multiplication by Fringe Shifting," *Experimental Mechanics,* Vol. 8, No. 12, 554-560 (Dec. 1968).
34. Post, D., "Analysis of Moiré Fringe Multiplication Phenomena," *Appl. Opt.,* Vol. 6, No. 11, 1938-1942 (1967).
35. Post, D., "New Optical Methods of Moiré Fringe Multiplication," *Experimental Mechanics,* Vol. 8, No. 2, 63-68 (Feb. 1968).
36. Post, D., "Moiré Fringe Multiplication with a Non-Symmetrical Doubly-Blazed Reference Grating," *Appl. Opt.,* Vol. 10, No. 4, 901-907 (April 1971).
37. Post, D. and MacLaughlin, T.F., "Strain Analysis by Moiré-Fringe Multiplication," *Experimental Mechanics,* Vol. 11, No. 9, 408-423 (Sept. 1971).
38. Daniel, I.M., Rowlands, R.E. and Post, D., "Strain Analysis of Composites by Moiré Methods," *Experimental Mechanics,* Vol. 13, No. 6, 246-252 (June 1973).
39. Wadsworth, N.J., Billing, B.F. and Marchant, M.J.N., "The Measurement of Local In-Plane Surface Displacement Using a Moiré Technique," Rep. No. RAE-TR-72046, Royal Aircraft Establishment, Farnborough, England, (April 1972).
40. Durelli, A.J. and Parks, V.J., *Moiré Analysis of Strain,* Prentice-Hall, Englewood Cliffs, NJ, 251-253 (1970).
41. Duncan, J.P. and Sabin, P.G., "An Experimental Method for Recording Curvature Contours in Flexed Elastic Plates," *Experimental Mechanics,* Vol. 5, No. 1, 22-28 (Jan. 1965).
42. Durelli, A.J., "Visual Representation of Kinematics of the Continuum," *Experimental Mechanics,* Vol. 6, No. 3, 113-139 (March 1966).
43. Lekhnitskii, S.G., *Theory of Elasticity of an Anisotropic Elastic Body,* Holden-Day, Inc., San Francisco, 163-174 (1963).
44. Savin, G.N., *Stress Concentrations Around Holes,* Pergamon Press, London, 152-174 (1961).

45. Chiang, F.P. and Slepetz, J., "Crack Length Measurements in Composites," *Journal of Composite Materials,* Vol. 7, 134-137 (1973).
46. Pipes, R.B. and Daniel, I.M., "Moiré Analysis of the Interlaminar Shear Edge Effect in Laminated Composites," *Journal of Composite Materials,* Vol. 5, 225-259 (1971).
47. Pipes, R.B. and Pagano, N.J., "Interlaminar Stresses in Composite Laminates Under Uniform Axial Extension," *Journal of Composite Materials,* Vol. 4, 538-548 (1970).
48. Chiang, F.P., Oplinger, D., Parker, B. and Slepetz, J., Discussion of "Strain Analysis of Composites by Moiré Methods" [by I.M. Daniel, R.E. Rowlands and D. Post, *Experimental Mechanics,* Vol. 13, No. 6, 246-252 (June 1973)], *Experimental Mechanics,* Vol. 14, No. 5, 206-207 (May 1974).
49. Samuelson, L.A., Vestergren, P., Knutsson, L., Wangberg, K.G. and Gamziukas, V., "Stability and Ultimate Strength of Carbon Fiber Reinforced Plastic Panels," *Advances in Composite Materials,* Vol. 1, ICCM 3, Third International Conf. on Composite Materials, Paris, France, 327-341 (Aug. 1980).
50. Rhodes, J.E., "Post-Buckling and Membrane Structural Capability of Composite Shell Structures," *Advances in Composite Materials,* Vol. 2, ICCM 3, Third International Conf. on Composite Materials, Paris, France, 1707-1720 (Aug. 1980).
51. D'Agostino, I., Drucker, D.C., Liu, C.K. and Mylonas, C., "An Analysis of Plastic Behavior of Metals with Bonded Birefringent Plastic," *Proceedings of the Society for Experimental Stress Analysis,* Vol. 12, 115-122 (1955).
52. Zandman, F. and Wood, M.R., "Photostress, A New Technique for Photo-Elastic Stress Analysis for Observing and Measuring Surface Strains on Actual Structures and Parts," *Product Engineering,* 167-178 (Sept. 1956).
53. Post, D. and Zandman, F., "Accuracy of Birefringent Coating Method for Coatings of Arbitrary Thickness," *Experimental Mechanics,* Vol. 1, No. 1, 21-32 (Jan. 1961).
54. Zandman, F., Redner, S.S. and Riegner, E.I., "Reinforcing Effect of Birefringent Coatings," *Experimental Mechanics,* Vol. 2, No. 2, 55-64 (Feb. 1962).
55. Redner, S.S., "New Oblique Incidence Methods for Direct Photoelastic Measurement of Principal Strains," *Experimental Mechanics,* Vol. 3, No. 3, 67-72 (March 1963).
56. Dally, J.W. and Alfirevich, I., "Application of Birefringent Coatings to Glass-Fiber-Reinforced Plastics," *Experimental Mechanics,* Vol. 9, No. 3, 97-102 (March 1969).
57. Daniel, I.M. and Rowlands, R.E., "Experimental Stress Analysis of Composite Materials," ASME Paper No. 72-DE-6, presented at ASME Design Engineering Conference, Chicago, IL (1972).
58. Yeow, Y.T., Morris, D.H. and Brinson, H.F., "The Fracture Behavior of Graphite/Epoxy Laminates," *Experimental Mechanics,* Vol. 19, No. 1, 1-8 (Jan. 1979).
59. Pipes, R.B. and Dalley, J.W., "On the Birefringent Coating Method of Stress Analysis for Fiber Reinforced Composite Laminates," *Experimental*

Mechanics, Vol. 12, No. 6, 272-277 (June 1972).
60. Pagano, N.J. and Pipes, R.B., "The Influence of Stacking Sequence on Laminate Strength," *Journal of Composite Materials,* Vol. 5, 50-57 (Jan. 1971).
61. Rybicki, E.F. and Hopper, A.T., "Analytical Investigation of Stress Con-Centrations Due to Holes in Fiber-Reinforced Plastic Laminated Plates, Three-Dimensional Models," Air Force Materials Lab. Report, AFML-TR-73-100 (June 1973).
62. Mandell, J.F., Wang, S.S. and McGarry, F.J., "The Extension of Crack Tip Damage Zones in Fiber Reinforced Plastic Laminates," *Journal of Composite Materials,* Vol. 9, 266-287 (June 1975).
63. Rowlands, R.E. and Daniel, I.M., "Application of Holography to Anisotropic Composite Plates," *Experimental Mechanics,* Vol. 12, No. 2, 75-82 (Feb. 1972).
64. Rowlands, R.E., Liber, T., Daniel, I.M. and Rose, P.G., "Stress Analysis of Anisotropic Laminated Plates," *AIAA Journal,* Vol. 12, No. 7, 903-908 (July 1974).
65. Maddux, G.E., "Photomechanics Facility," Air Force Flight Dynamics Laboratory (1976).
66. Rowlands, R.E., Liber, T., Daniel, I.M. and Rose, P.G., "Higher-Order Numerical Differentiation of Experimental Information," *Experimental Mechanics,* Vol. 13, No. 3, 105-112 (March 1973).
67. Stetson, K.A., "Moiré Method for Determining Bending Moments from Hologram Interferometry," *Optics Technology,* Vol. 2, 80-84 (1970).
68. Powell, R.L. and Stetson, K.A., "Interferometric Vibration Analysis by Wave Front Reconstruction," *J. Optical Soc. America,* Vol. 55, 1593-1598 (1965).
69. Sampson, R.C., "Holographic Interferometry Applications in Experimental Mechanics," *Experimental Mechanics,* Vol. 10, No. 8, 313-320 (Aug. 1970).
70. Wells, D.R., "NDT of Sandwich Structures by Holographic Interferometry," *Materials Evaluation,* Vol. 27, 225 (1969).
71. Leith, E.N. and Vest, C.M., "Investigation of Holographic Techniques," Report No. 2420-9-P, Willow Run Laboratories, The University of Michigan, Ann Arbor, MI (April 1970).
72. Iversen, R.J., Schulz, R.D. and Arnold, R.G., "Holographic Testing Techniques," Report No. AMSWE-RE-71-32, Army Weapons Command, Rock Island, IL (June 1971).
73. Erdmann-Jesnitzer, F. and Winkler, T., "Application of the Holographic Nondestructive Testing Method for Evaluation of Disbonding in Sandwich Plates," *Advances in Composite Materials,* Vol. 2, ICCM 3, Third International Conf. on Composite Materials, Paris, France, 1029-1039 (Aug. 1980).
74. Erf, R.K., Waters, J.P., Gagosz, R.M., Michael, F. and Whitney, G., "Nondestructive Holographic Techniques for Structures Inspection," Tech. Report AFML-TR-72-204 (Oct. 1972).
75. Sendeckyj, G.P., Maddux, G.E. and Tracy, N.A., "Comparison of Holographic, Radiographic and Ultrasonic Techniques for Damage Detection in Composite Materials," *ICCM/2, Proceedings of the 1978 International Con-*

ference on Composite Materials, Toronto, Canada (April 16-20, 1978); Metallurgical Society of AIME, 1037-1056.

76. Maddux, G.E. and Sendeckyj, G.P., "Holographic Techniques for Defect Detection in Composite Materials," *Nondestructive Evaluation and Flaw Criticality for Composite Materials,* ASTM STP 696, R.B. Pipes, ed., American Society for Testing and Materials, 26-44 (1979).
77. Freund, N.P., "Measurement of Thermal and Mechanical Properties of Graphite/Epoxy Composites for Precision Applications," *Composite Reliability,* ASTM STP 580, American Society for Testing and Materials, 133-145 (1975).
78. Leendertz, J.A., "Interferometric Displacement Measurement on Scattering Surfaces Utilizing Speckle Effect," *Journal Physics E. (J. Scientific Instruments),* Vol. 3, 214 (1970).
79. Archbold, E., Burch, J.M. and Ennos, A.E., "Recording of In-Plane Surface Displacement by Double-Exposure Speckle Photography," *Optica Acta,* Vol. 17, No. 12, 883-898 (1970).
80. Hung, Y.Y. and Hovanesian, J.D., "Full-Field Surface-Strain and Displacement Analysis of Three-Dimensional Objects by Speckle Interferometry," *Experimental Mechanics,* Vol. 12, No. 10, 454-460 (Oct. 1972).
81. Hung, Y.Y. and Taylor, C.E., "Speckle-Shearing Interferometric Camera—A Tool for Measurement of Derivatives of Surface Displacement," *Proceedings of the Society of Photo-Optical Instrumentation Engineers,* Vol. 41, San Diego, CA, 169-176 (Aug. 27-29, 1973).
82. Hung, Y.Y., Rowlands, R.E. and Daniel, I.M., "Speckle-Shearing Interferometric Technique: A Full-Field Strain Gage," *Applied Optics,* Vol. 14, No. 3, 618-622 (March 1975).
83. Hung, Y.Y., Daniel, I.M. and Rowlands, R.E., "Full-Field Optical Strain Measurement Having Postrecording Sensitivity and Direction Selectivity," *Experimental Mechanics,* Vol. 18, No. 2, 56-60 (Feb. 1978).
84. Hung, Y.Y. and Hovanesian, J.D., "Nondestructive Testing by Speckle-Shearing Interferometry," *Proc. of 12th Symposium on Nondestructive Evaluation,* San Antonio, TX, 163-167 (April 24-26, 1979).
85. Pih, H. and Knight, C.E., "Photoelastic Analysis of Anisotropic Fiber Reinforced Composites," *J. Composite Materials,* Vol. 3, 94-107 (Jan. 1969).
86. Sampson, R.C., "A Stress-Optic Law for Photoelastic Analysis of Orthotropic Composites," *Experimental Mechanics,* Vol. 10, No. 5, 210-215 (May 1970).
87. Dally, J.W. and Prabhakaran, R., "Photo-orthotropic Elasticity," *Experimental Mechanics,* Vol. 11, No. 8, 346-356 (Aug. 1971).
88. Bert, C.W., "Theory of Photoelasticity of Birefringent Filamentary Composites," *Fibre Science and Technology,* Vol. 5, 165-171 (1972).
89. Pipes, R.B. and Rose, J.L., "Strain-Optic Law for a Certain Class of Bi-Refringent Composites," *Experimental Mechanics,* Vol. 14, No. 9, 355-360, (Sept. 1974).
90. Prabhakaran, R., "On the Stress-Optic Law for Orthotropic Model Materials in Biaxial Stress Fields," *Experimental Mechanics,* Vol. 15, No. 1, 29-34 (Jan. 1975).

91. Cernosek, J., "Note on Photoelastic Response of Composites," *Experimental Mechanics,* Vol. 15, No. 9, 354-357 (Sept. 1975).
92. Knight, C.E. and Pih, H., "Orthotropic Stress-Optic Law for Plane Stress Photoelasticity of Composite Materials," *Fibre Science and Technology,* Vol. 9, 297-313 (1976).
93. Prabhakaran, R., "A Strain-Optic Law for Orthotropic Model Materials," *AIAA Journal,* Vol. 13, 723-728 (1975).

4 COMPOSITE CHARACTERIZATION

4.1 CONSTITUENT TEST METHODS

It is often desirable to obtain constituent properties as part of the total characterization of composite materials. Tensile properties of fiber and matrix are of particular interest. Determination of matrix properties may be hindered for certain resin systems by the inability to fabricate castings. This problem can usually be overcome, however, by utilizing thin film specimens. Many of the physical property characterization methods for constituents are essentially the same as for composites. As a result, these methods will be covered under paragraph 4.2.

4.1.1 Single Filament Tensile Properties

This test method is used to determine strength, modulus, and failure strain of single filaments. The method consists of applying constant-strain-rate loading to a single filament, center-line mounted on special slotted tabs (see Fig. 4-1). This method is discussed in detail in ASTM D3379-75, and is limited to fibers with a modulus above 21 × 10^9 Pa (3 × 10^6 psi). The fixed gage length of the fibers, L, must be at least 2000 times the nominal filament diameter. The gripping system used in conjunction with the tabs shown in Fig. 4-1 must be designed such that axial alignment may be easily accomplished without damaging the filament. After the specimen is mounted in the test machine, the center section of the tab is burned or cut away to allow for filament elongation.

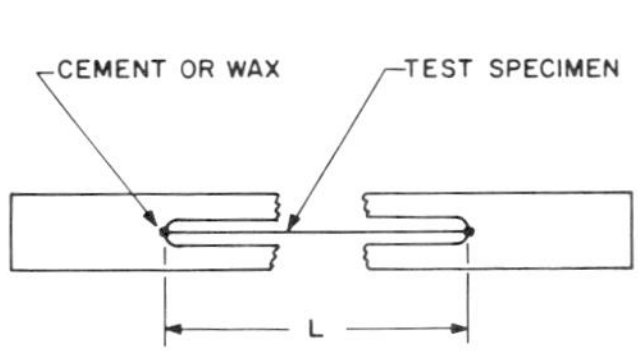

Fig. 4-1—Single filament mounting method, ASTM D 3379-75

Filament cross-sectional area, A, is determined by planimeter measurements of a representative number of filament cross sections as displayed on highly magnified photomicrographs. For fibers with large cross-sectional variations, such as graphite, a large number of sample measurements is necessary in order to obtain an accurate average value of the cross-sectional area. Other methods of cross-sectional area determination include optical gages and image-splitting microscope.

The test procedure consists of loading the specimen to failure at a constant cross-head rate and recording a load-displacement curve. The filament strength, σ_f, is simply the maximum load, P_{max}, divided by the average initial cross-sectional area, A

$$\sigma_f = \frac{P_{max}}{A} \tag{4.1}$$

An apparent compliance, C_a, can be determined from the initial straight line portion of the load-displacement curve, i.e.,

$$C_a = \frac{u}{P} \tag{4.2}$$

where P and u are the load and cross-head displacement, respectively, associated with the slope of the initial straight line portion of the load-displacement curve on the recording chart. The term "apparent" compliance is used because cross-head travel, due to system compliance, is not a true measure of filament displacement. This difficulty can be overcome by assuming the displacement, u_s, associated with system compliance is constant for a particular fiber and tab. Thus

$$\frac{u}{P} = \frac{u_f}{P} + \frac{u_s}{P} = \frac{1}{AE_f} L + \frac{u_s}{P} \tag{4.3}$$

where u_f and E_f are the actual fiber displacement and modulus, respectively. If values of C_a are determined for different gage length specimens at a constant value of P, then a true value of E_f can be determined from eq (4.3). In particular, E_f can be determined from the slope of a best-fit straight line to a plot of measured values of C_a as a

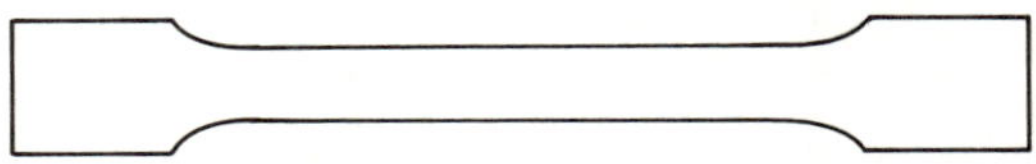

Fig. 4-2—Dogbone specimen for polymeric matrix tensile properties, ASTM D638-72

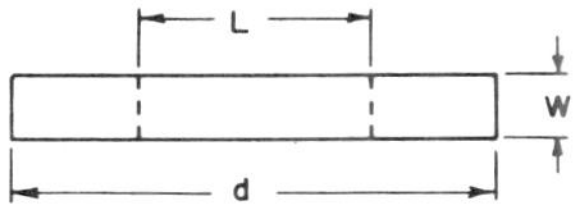

d - L ≥ 50 mm (2 in)
5 mm (0.2 in) ≤ W ≤ 25.4 mm (1.0 in)
W/h ≥ 8

Fig. 4-3—Thin strip specimen for polymeric matrix tensile properties, ASTM D882-73

function of L. The y-intercept of this straight line is the machine compliance, C_s, from which u_s can be determined.

$$u_s = C_s P \tag{4.4}$$

The fiber strain at failure, ϵ_f, is simply

$$\epsilon_f = \frac{u_{max} - u_s}{L} \tag{4.5}$$

where u_{max} is the maximum cross-head displacement as determined from the recording chart.

A more accurate strain measurement can be obtained by optical techniques.

4.1.2 Polymeric Matrix Tensile Properties

For resin systems which can easily be processed in thick sheets, a standard dogbone type specimen, as shown in Fig. 4-2, is used for tensile property determination. The exact dimensions for the dogbone specimen depend on thickness. Details are spelled out is ASTM D638-72. The specimens may be prepared by machining or die cutting from materials in sheet or plate form, or they may be prepared by molding.

Strain measurements can be obtained by using an extensometer. If a measure of Poisson's ratio is desirable, transverse strain gages can be employed, provided the specimen is sufficiently thick and that a correction for gage transverse sensitivities (see section 3.1.3) is applied.

Resin systems which cannot be readily fabricated in thick sheets can often be fabricated into thin sheets or films. Tensile tests can be performed on strips of such material cut from sheets or molded. The details of this test method are discussed in ASTM D882-73. The method is applicaple to specimens with a thickness, h, less than 1.0 mm (0.04 in.). The specimen is shown in Fig. 4-3 where the test gage section is denoted by L. For specimens which are not too thin, an extensometer can be used to measure strain. For flexible specimens, optical techniques may be used for strain measurement.

It should also be noted that a dogbone-type film specimen can be utilized by molding to the desired dimensions.

4.2 PHYSICAL PROPERTY TEST METHODS

In this section, measurement of density, fiber volume content, and expansional strains are discussed. The method for density measurement is applicable to constituent materials as well as composites, while methods for determining thermal and moisture expansion coefficients are applicable to polymeric matrix materials and composites.

4.2.1 Density

Density for polymeric matrix resins and fiber reinforced composites may be determined by measuring the difference between the weight of a specimen in air and in water. The volume of the specimen must be at least 1 cm^3 (0.06 $in.^3$), and the surface edges should be smooth.

The test procedure consists of measuring the specimen weight in air, then weighing the specimen while suspended on wire and immersed in a container of water. For specimens with a specific gravity less than unity, a sinker may be attached to the wire. It is also necessary to measure the weight of the completely immersed sinker and partially immersed wire. The density, ϱ, is then determined from the relationship

$$\varrho = \frac{(0.9975)a}{(a + w - b)} \tag{4.6}$$

where a is the apparent weight of the specimen in air, b is the apparent weight of specimen and sinker completely immersed and of the wire partially immersed, and w is the weight of the totally immersed sinker and partially immersed wire. This test method is covered by ASTM D792-66.

A similar procedure may be used for measuring density of single filaments. The liquid chosen, however, must be capable of completely wetting the fiber.

4.2.2 Fiber Volume Fraction

Two methods are available for determining fiber volume content. One method involves digestion of the matrix by a liquid medium with the weight of the remaining fiber being used in conjunction with the fiber and composite densities to determine volume percent of fiber. A second method involves a simple determination of the composite density. If the composite is essentially free of voids, then the composite density can be used in conjunction with the fiber matrix densities to calculate fiber volume content.

The digestion method is applicable to both polymeric and metal matrix composites. Choice of liquid for digestion is governed by the specific matrix and fiber. For example, epoxy resins are readily digested by sulfuric acid, but graphite fibers are attacked by sulfuric acid. As a result, hot nitric acid is used in conjunction with graphite/epoxy composites. Nitric acid will also attack graphite if allowed to remain in the acid for too long a time period. The attack, however, is much less severe than in the case of sulfuric acid. Thus, the choice of liquid for digestion must be compatible with the fiber while doing an efficient job of digesting the matrix. More details concerning the choice of liquid for digestion can be found in ASTM D3171-76 for polymeric matrix composites, and in ASTM D3553-76 for metal matrix composites.

The test procedure consists of weighing the composite sample and placing it in the hot liquid medium until the matrix is dissolved. The residue is then filtered, washed, dried, and weighed. Volume fraction of fiber, V_f, is then determined from the relationship

$$V_f = \frac{(W_f/\varrho_f)}{(W_c/\varrho_c)} \tag{4.7}$$

where W_f, W_c, ϱ_f, and ϱ_c are the fiber weight, composite weight, fiber density, and composite density, respectively. If the matrix density, ϱ_m, is known, the void volume fraction, V_v, can be calculated from the relationship

$$V_v = 1 - \left[\frac{W_f/\varrho_f + (W_c - W_f)/\varrho_m}{W_c/\varrho_c}\right] \tag{4.8}$$

If some weight loss by the fiber is suspected, a correction can be made by exposing a bundle of fibers (of approximately the same weight as the fibers in the composite sample) to the liquid medium for the same length of time as the composite exposure. Fiber weight loss is then determined by measuring the difference between the fiber weight before and after exposure. Only weight losses greater than 0.5 percent are considered significant. Caution should be exercised on this correction procedure, however, as the fibers in the composite are not exposed to direct contact with the digesting medium during the full time that the resin is being dissolved.

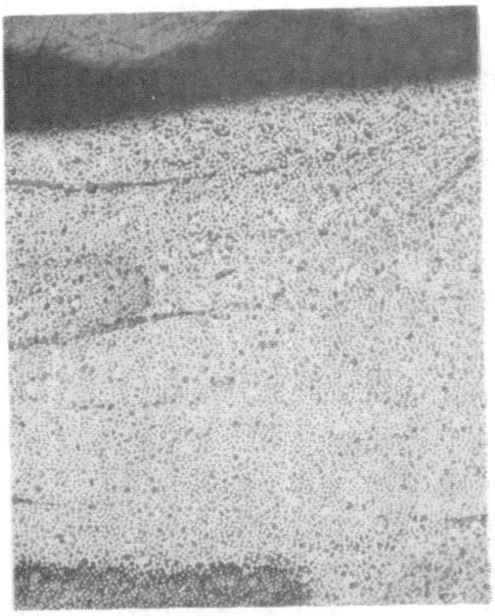

(A)

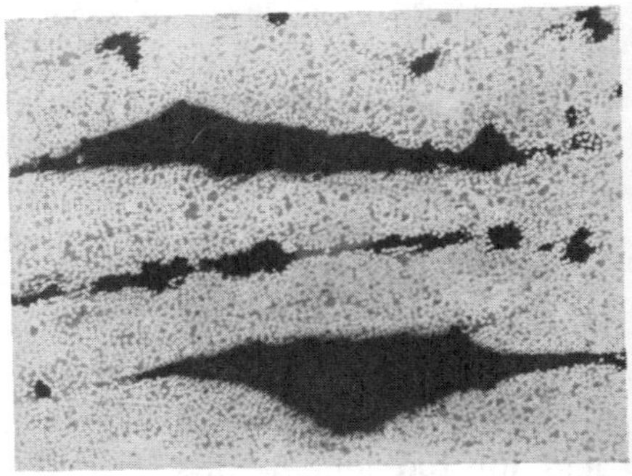

(B)

Fig. 4-4—Unidirectional graphite/epoxy composite cross-sections: (a) no void content, (b) large void content

Because of the extreme pressure used in processing metal matrix composites, they are essentially void-free. In addition, many polymeric matrix composites are fabricated with void contents less than one percent. In such cases it is assumed that the void content is zero and the fiber volume fraction can be determined from the densities of the fiber, composite, and matrix through the relationship

$$V_f = \left(\frac{\varrho_c - \varrho_m}{\varrho_f - \varrho_m}\right) \tag{4.9}$$

Void content can be checked from photomicrographs of composite cross-sections. Typical cross-sections are shown in Fig. 4-4 for a graphite/epoxy unidirectional composite. The cross section in Fig. 4-4(a) shows a large void content (dark spots are voids), while the cross section in Fig. 4-4(b) is void-free.

It should be noted that many fibers, such as graphite, tend to have a large variation in density. In such cases, both of the procedures discussed for determining fiber volume content yield only approximate results.

4.2.3 Coefficient of Thermal Expansion

Measurement of linear coefficient of thermal expansion simply involves measuring the dimensional change of a specimen over a temperature range. To verify linearity, it is necessary to plot thermal strain as a function of temperature. Typical results are shown in Fig. 4-5 for

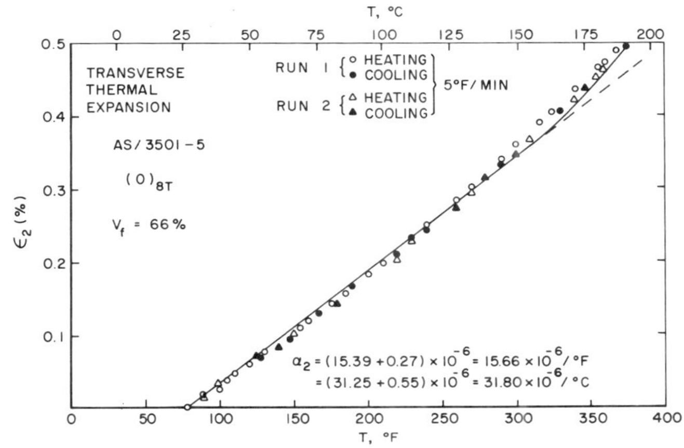

Fig. 4-5—Transverse thermal expansional strain as a function of temperature for AS/3501-5 unidirectional graphite/epoxy composite (data, courtesy of R. Kim, University of Dayton Research Institute)

the transverse thermal expansion coefficient, α_2, of a Hercules' AS/3501-5 unidirectional graphite/epoxy composite.[1] Transverse strain, ϵ_2, was measured using a strain gage. The linear coefficient of thermal expansion is the slope of the ϵ_2 versus T plot over the initial straight line portion of the curve. A factor is added to compensate for the difference between thermal expansion coefficient of the gage and the composite. For the example in Fig. 4-5, the curve is linear up to a temperature which is close to the composite cure temperature, where the material response displays considerable viscoelasticity.

Other appropriate devices, such as an extensometer, may be used to measure thermal strain.

This technique can also be used on polymeric resins which can be fabricated into thick sheets. A detailed procedure for cast resins can be found in ASTM D696-70.

Since moisture induces swelling in many resin systems, the specimens used for determining coefficient of thermal expansion should be dry initially.

4.2.4 Coefficient of Moisture Expansion

Measurement of coefficient of moisture expansion involves measuring the dimensional change of a specimen over a range of moisture weight gains. The experiment may be performed by placing a specimen in a water bath and measuring swelling strain as a function of weight

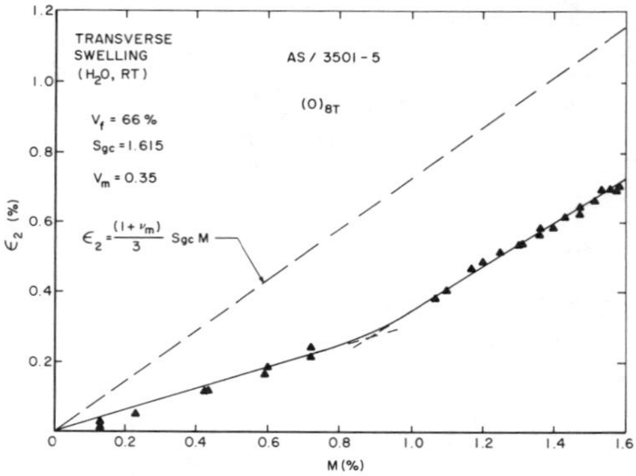

Fig. 4-6—Transverse swelling strain as a function of moisture content for AS/3501-5 graphite/epoxy composite (Hahn and Kim[1])

gain. Difficulty arises, however, in the determination of strain, as conventional strain gage adhesives are attacked by moisture. Thus, the accuracy of strain gage measurements is reduced with exposure time.

Hahn and Kim[1] used a micrometer and a caliper to measure swelling strains on graphite/epoxy composites. The gage length of the micrometer was 2.54 cm with a resolution of 0.00254 mm, and the gage length of the caliper was 20.3 cm with a resolution of 0.0127 mm. Strain measurements were based on an average of three readings for inplane strains and an average of four readings for through-the-thickness strains. This technique can be used on both neat resins and composites.

In the case of unidirectional composites, most of the swelling occurs transverse to the fibers, while most of the swelling occurs through-the-thickness for laminated composites.

It has been noted by Hahn[2] that moisture absorption in graphite/epoxy composites is associated with a swelling threshold described by the relationship

$$\begin{aligned} \epsilon_i^H &= 0, \qquad c \leq c_o \\ \epsilon_i^H &= \beta_i(c - c_o), \qquad c_o < c \end{aligned} \tag{4.10}$$

where c is moisture concentration in general and c_o is a threshold moisture concentration. Such a threshold moisture level may be associated with water filling micro-voids or cracks.

Because of the relationship given by eq (4.10) care must be exercised in measuring β_1 from a plot of swelling strain versus total weight gain.

Consider the case of a unidirectional graphite/epoxy composite in which the inplane dimensions are large compared to the thickness such that the moisture diffusion is essentially one-dimensional in nature. A plot of the average transverse strain, ϵ_2, versus total percent weight gain, M, yields the result shown in Fig. 4-6. The initial portion of this curve is associated with a moisture gradient which has not exceeded c_o throughout the laminate. At some time, $t = t_o$, $c \geq c_o$ throughout the composite and the swelling strains will be linear with the total weight gain. This can be illustrated using eq (4.10) as follows

$$\epsilon_2(t) = \frac{2}{h}\int_o^{h/2} \epsilon_2^H dz = \beta_2 \varrho_c[M(t) - M_o] \qquad (4.11)$$

$$t_o \leq t$$

where ϵ_2 denotes the average strain through the thickness as measured in the experiment and M_o denotes the weight gain associated with the threshold concentration c_o. Thus, if S_2 is the slope of the secondary straight line portion of Fig. 4-6

$$\beta_2 = \frac{S_2}{\varrho_2} \qquad (4.12)$$

The value of M_o can be obtained by extending the secondary straight line portion of the curve through the M-axis.

The dotted line in Fig. 4-6 is based on the assumption that all of the absorbed moisture is translated into a matrix volume change. Such behavior does occur after the threshold moisture gain has been attained throughout the composite. Using micromechanics[2]

$$\epsilon_2(t) = \frac{(1 + \nu_m)}{3} S_{gc}[M(t) - M_o], \qquad (4.13)$$

$$t_o \leq t$$

where S_{gc} is the specific gravity of the composite. Combining eqs (4.11) and (4.13) yields

$$\beta_2 = \frac{(1 + \nu_m)}{3\varrho_w} \qquad (4.14)$$

where ϱ_w is the density of water.

For some composites and neat resins, c_o may vanish and the swelling strains will be linear with M up to full saturation.

4.3 COMPOSITE TEST METHODS

Characterization of composite material systems requires that experimental tests be performed which yield *intrinsic* strength and stiffness properties of the material. Much effort has been expended in developing test methods which are appropriate in assessing material variations due to processing or raw material variables. However, these "quality assurance" test methods do not generally yield data which may be directly related to material performance. Although the test methods reviewed in this section were designed to provide intrinsic material property data, the degree of achievement of this goal varies from method to method. Therefore, as each test method is discussed in the following paragraphs, the nature of the material property data developed by a given test method will be discussed with respect to the desired intrinsic material property data.

The fundamental principle underlying test methods for laminated composites can be stated as follows:

> *The unidirectional lamina is the building block of the multidirectional laminate. Therefore, characterization of lamina material properties allows predictions of the properties of any laminate.*

In actual practice, the above statement has been demonstrated to hold only for prediction of laminate elastic stiffness properties such as effective modulus or Poisson's ratio. However, prediction of laminate strength properties from intrinsic lamina strength data has proven more difficult. Therefore, in certain cases it is necessary to resort to direct characterization of laminate strength properties. With this in mind, the following discussions will treat test methods for both lamina and laminate characterizations.

4.3.1 Tensile Test Methods

The tensile test is performed in order to determine uniaxial tensile strength, Young's modulus and Poisson's ratio. The tensile test method allows determination of the following properties for the *lamina*

E_1 = Young's modulus in the fiber direction
E_2 = Young's modulus transverse to the fiber direction
ν_{12} = major Poisson's ratio
ν_{21} = minor Poisson's ratio
X_1^T = ultimate tensile strength in fiber direction
X_2^T = ultimate tensile strength transverse to fiber direction
e_1^T = ultimate strain in the fiber direction
e_2^T = ultimate strain transverse to fiber direction

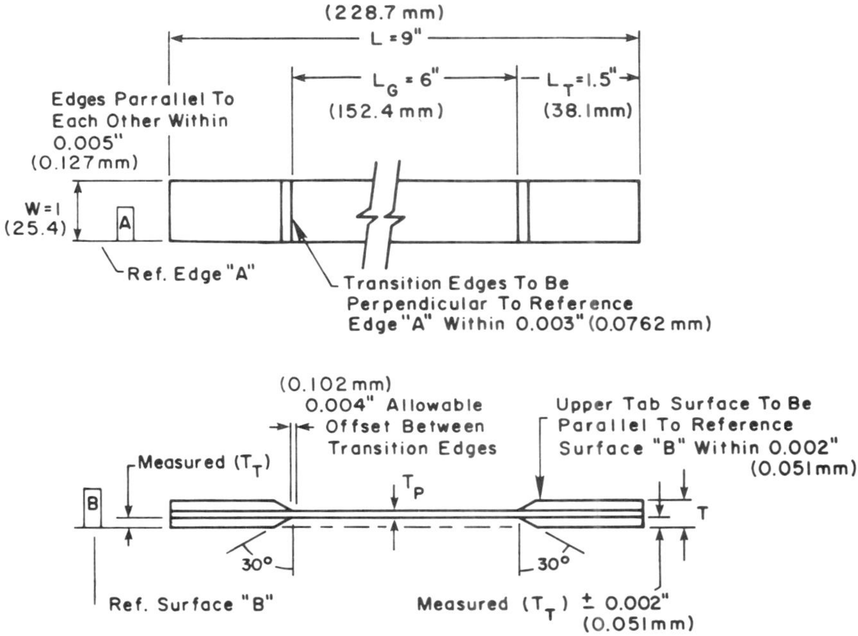

W	L	L_G	L_T	T	T_T	T_P	
±0.0156	±0.1250	±0.0625	±0.0625	±0.0150	±0.0050	——	inches
0.396	3.175	1.5875	1.5875	0.381	0.127		mm

Fig. 4-7—Geometry of tensile coupon

The tensile specimen is straight-sided and of constant cross-section with adhesively bonded and beveled tabs for load introduction. The lamina 0-deg test specimen is 0.500 inches in width, and six plies in thickness while the lamina 90-deg test specimen is 1.000 inches in width and eight plies in thickness. The overall length of the specimen is 9.0 inches and the test section is 6.0 inches. The test specimen geometry is illustrated in Fig. 4-7. Test specimen thickness variation should not exceed ±2 percent. Specimens should be precision machined from plates with tabs bonded in place. Test specimen edges should be

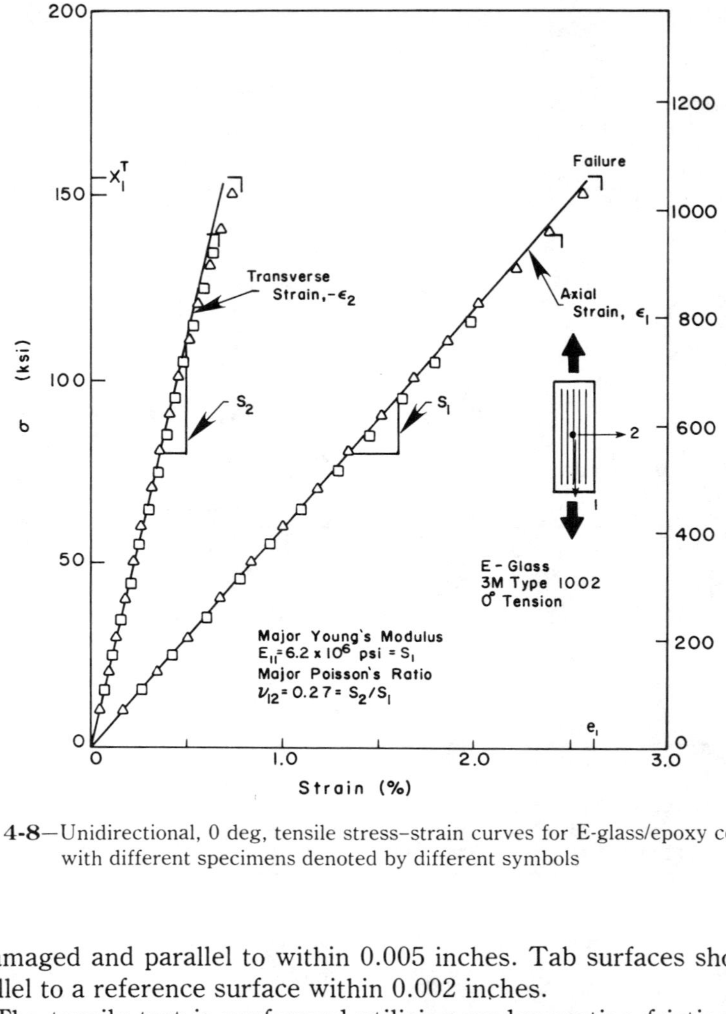

Fig. 4-8—Unidirectional, 0 deg, tensile stress–strain curves for E-glass/epoxy composite with different specimens denoted by different symbols

undamaged and parallel to within 0.005 inches. Tab surfaces should be parallel to a reference surface within 0.002 inches.

The tensile test is performed utilizing wedge-section friction grips. The specimen is first aligned in the grips and tightened in place. The specimen is then loaded monotonically to failure at a recommended rate of 0.02 cm/min. In order to determine specimen strains, electrical resistance strain gages are mounted on the specimen and monitored during the test. Foil gages with a resistance of 350 ohms and a gage length of 0.125 or 0.250 inches have been found to minimize gage heating during the test. Typically, three gages are applied to each specimen, two longitudinal and one transverse to the load direction. The gages are located at the geometric center of the test section with the longitudinal gages mounted on opposite faces of the specimen. The

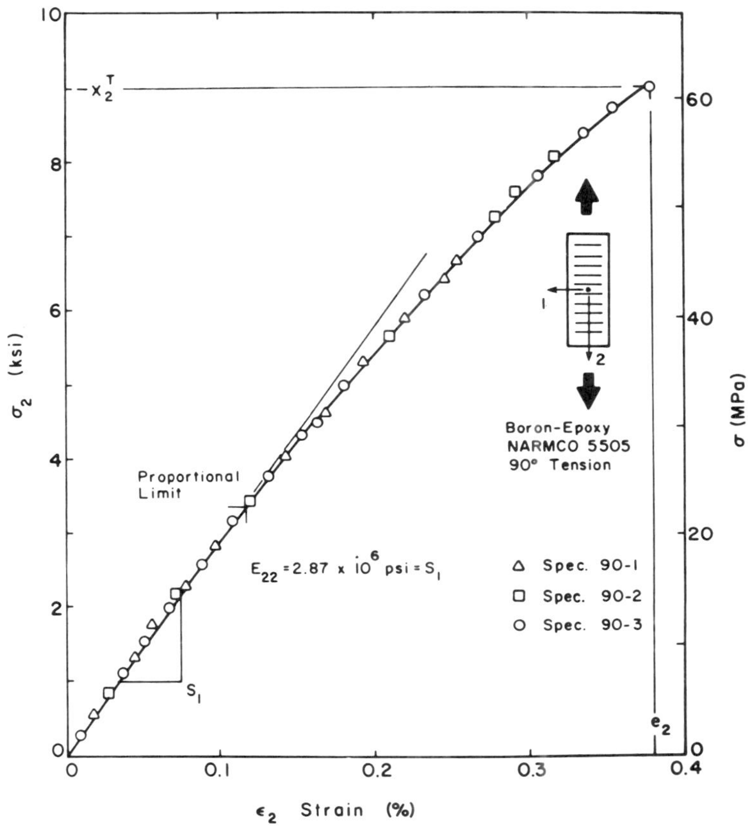

Fig. 4-9—Unidirectional, 90 deg, tensile stress–strain curve for boron/epoxy composite with different specimens denoted by different symbols

gages may be monitored continuously or at discrete load intervals. If discrete data are taken, sufficient data points to establish modulus must be taken within the region of linear elastic behavior. Typical test results for the 0-deg tensile test are shown in Fig. 4-8 for an *E*-glass/epoxy composite. Test results for the 90-deg tensile test of a boron–epoxy composite are shown in Fig. 4-9. The relevant test method is ASTM D3039-76.

Tensile tests of *orthotropic laminate* ($A_{16} = A_{26} = 0$) specimens of less than 0.100 inches in thickness are accomplished with a specimen geometry and test procedure identical to that discussed previously for the lamina 90-deg tensile tests. Such a test will allow determination of the following laminate "effective" properties:

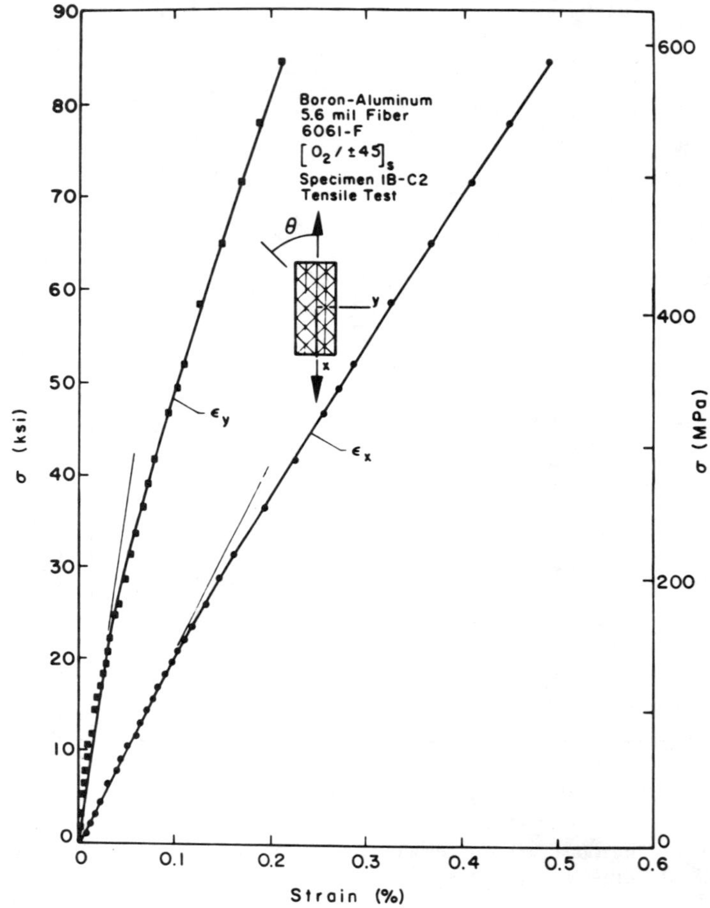

Fig. 4-10—Tensile stress–strain curves for boron/aluminum $[0_2/\pm45]_s$ laminate

E_x = effective laminate modulus in loading direction
ν_{xy} = effective laminate Poisson's ratio
$\overline{X}^T$ = laminate uniaxial ultimate stress
$\bar{e}^T$ = laminate uniaxial ultimate strain

Typical laminate tensile test results for the $[0_2/\pm45]_s$ boron/aluminum laminate are shown in Fig. 4-10. Note that it is necessary to measure discrete stress–strain results at relatively low stress levels in order to establish the onset of the aluminum matrix yield.

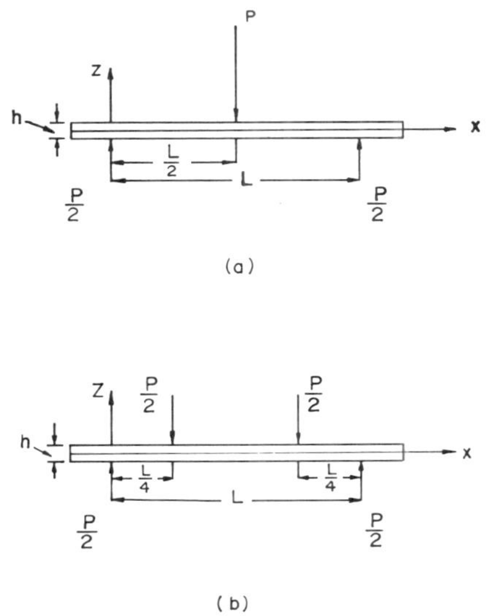

Fig. 4-11—Flexure test: (a) three-point loading; (b) four-point loading

4.3.2 Flexure Test Methods

The flexure test is based on the loading configuration shown in Fig. 4-11, where the four-point flexure test is shown in Fig. 4-11(a) and the three-point flexure test in Fig. 4-11(b). It should be noted that the four-point flexure test is sometimes utilized with the load points at $L/3$ as well as at quarter points as shown. The quarter-point loading has been utilized primarily for high modulus materials such as graphite/epoxy and boron/epoxy. These test methods are used for determining flexural strength and modulus. The test is not recommended for generating design data, but it does provide a simple test for quality control. If proper interpretation is given to the data, however, it may have more value than a quality control tool. This will be discussed in detail in the next section.

The flexure methods are applicable to polymeric and composite materials which can be treated as homogeneous. For laminated materials the results will depend on ply stacking sequence, requiring the data to be interpreted in conjunction with laminated beam theory.[3] Because of this complexity, flexure tests in conjunction with high modulus, continuous filament composites are generally limited to unidirectional materials with fibers oriented at either 0 deg or 90 deg to the beam axis.

A testing machine having a controllable crosshead speed is used in conjunction with a loading fixture, such as the one shown in Fig. 4-12 for a four-point flexure test. Specific requirements for radius of the

Fig. 4-12—Typical loading fixture used for four-point flexure test

load noses and supports are given in ASTM D790-71. Span-to-depth ratios, L/h, depend on the ratio of tensile strength parallel to the beam axis to interlaminar shear strength. For strength ratios less than 8 to 1, an L/h ratio of 16 is recommended. This is typical of fiberglass composites. For high modulus materials such as graphite/epoxy and boron/epoxy a value of $L/h = 32$ is recommended for 0-deg unidirectional composites. For 90-deg unidirectional properties of these materials, $L/h = 16$ is appropriate. Recommended specimen dimensions can also be found in ASTM D790-71. Crosshead travel rate, R, should be chosen such that the strain rate of the outer surface, $\dot{\epsilon}$, is approximately 0.01 mm/mm/min (0.01 in./in./min). For both the three-point flexure and four-point flexure at quarter points,

$$\dot{R} = \frac{\dot{\epsilon} L^2}{6h} \tag{4.15}$$

For tensile strength determination the specimen is loaded until failure occurs. The maximum tensile stress, σ_m, is determined for the three-point loading from the relationship

$$\sigma_m = \frac{3PL}{2bh^2} \tag{4.16}$$

where b is the beam width. For four-point loading at quarter points

$$\sigma_m = \frac{3PL}{4bh^2} \tag{4.17}$$

It should be noted that eqs (4.16) and (4.17) are valid only for materials in which the stress–strain curve is linear to failure. If some nonlinear stress–strain behavior occurs, an error will be introduced into the relationships for σ_m. An adjustment in the calculation for σ_m must also be made if the deflections become large (see ASTM D790-71).

For the determination of modulus a deflectometer or an extensometer can be used to measure center deflection. If a deflectometer is used, the test must be stopped at a point where the stress–strain behavior is still linear and a reading taken. The use of an extensometer allows a continuous plot of load-deflection. Crosshead travel does not provide a very accurate measurement of deflection due to machine compliance. Although the error will be less for large strains, modulus is determined from the initial linear portion of the load-deflection curve where strains are usually small. Modulus is determined for the three-point flexure test from the relationship

$$E_x = \frac{PL^3}{4bh^3w}(1 + S) \tag{4.18}$$

where w is the center deflection and S is a correction factor for shear deformation given by

$$S = \frac{3h^2E_x}{2L^2G_{xz}} \tag{4.19}$$

and G_{xz} is the shear modulus of the composite in a longitudinal plane through the thickness. For four-point loading at quarter points

$$E_x = \frac{PL^3}{64bh^3w}(11 + 8S) \tag{4.20}$$

Note that the shear correction factor is a function of L/h and E_x/G_{xz}. For high modulus materials such as graphite/epoxy and boron/epoxy, $E_x/G_{xz} > 20$ for 0-deg unidirectional material. Thus, shear deformation will be significant even for a span-to-depth ratio of 32. In many cases, G_{xz} may not be known, but an approximate value of E_x/G_{xz} is known. If a value of G_{xz} can be determined, eqs (4.18) and (4.20) can be used in the form

$$E_x = \frac{PL^3}{4bh^3\left[w - \frac{3PL}{8bhG_{xz}}\right]} \tag{4.21}$$

for three-point loading and

$$E_x = \frac{11PL^3}{64bh^3\left[w - \frac{3PL}{16bhG_{xz}}\right]} \tag{4.22}$$

for four-point loading at quarter points. If shear deformation is neglected, eqs (4.18) and (4.20) take the simplified form

$$E_x = \frac{PL^3}{4bh^3w} \tag{4.23}$$

for three-point loading and

$$E_x = \frac{11PL}{64bh^3w} \tag{4.24}$$

A cursory examination of eqs (4.18) and (4.19) reveals that the neglect of shear deformation yields an apparent modulus which is less than the actual value of E_x. Thus, if the shear correction factor, S, cannot be easily determined, the flexure tests can be run for modulus at increasing ratios of L/h until a constant value of E_x is measured. For high values of L/h, however, large deflections may occur at low stress levels. Thus, the applied load should be just sufficient to provide an initial slope of the load-deflection curve. Obviously, it is also necessary to keep loads below specimen damage levels.

The four-point load test is often preferred to the three-point method because the center section is under pure bending stresses. Shear stresses do exist, however, between the outer supports and the applied load. The maximum shear stress, τ_m, is given by

$$\tau_m = \frac{h}{2L}\sigma_m \tag{4.25}$$

for three-point loading and by

$$\tau_m = \frac{h}{L}\sigma_m \tag{4.26}$$

for four-point loading at quarter points. Thus, for fixed values of L/h the maximum shear stress in the four-point test is twice the maximum shear stress in the three-point loading. Equations (4.25) and (4.26) can be used to determine an L/h ratio which will assure tensile failure. When flexure tests are run at elevated temperatures, the shear strength may drop significantly causing shear failures. As a result, larger span-to-depth ratios may be necessary to produce elevated temperature tensile failures.

4.3.3 Comparison Between Tensile and Flexure Strength

Tensile data generated from a unidirectional flexure test usually yield higher strength than data obtained from a standard tensile coupon. It is primarily for this reason that flexure data are not considered appropriate for design purposes. This difference in apparent tensile strengths can be accounted for, however, if one considers the brittle nature of most polymeric matrix composites. In particular, the Weibull statistical strength theory for brittle materials[4] states that the probability of survival, P, at a stress level, σ, for a uniaxial stress field in a homogeneous material, governed by a volumetric flaw distribution is given by

$$P(\sigma) = \exp[-B(\sigma)] \tag{4.27}$$

where B is the risk of rupture. For a two-parameter Weibull model, which has been used in conjunction with composite materials,[3-5,6,7] the risk of rupture is given by the volume integral

$$B(\sigma) = \int_v \left(\frac{\sigma}{\hat{\sigma}}\right)^\alpha dV \tag{4.28}$$

where $\hat{\sigma}$ is a location parameter of the distribution, often referred to as the characteristic strength, and α is the shape parameter. Both of these parameters are considered to be material properties independent of size. Thus, the risk to break will be a function of the stress distribution in the test specimen. For the case of a simple tensile test under uniform stress, eq (4.28) becomes

$$B_t = V_t \left(\frac{\sigma_t}{\hat{\sigma}}\right)^\alpha \tag{4.29}$$

where the subscript t denotes simple tension. Equation (4.27) can now be written in the form

$$P(\sigma_t) = \exp\left[-\left(\frac{\sigma_t}{\hat{\sigma}_t}\right)^{\alpha}\right] \tag{4.30}$$

where $\hat{\sigma}_t$ is the location parameter for tensile loading given by

$$\hat{\sigma}_t = \frac{\hat{\sigma}}{(V_t)^{1/\alpha}} \tag{4.31}$$

Thus, tensile tests from specimens of different dimensions could be represented by a two-parameter Weibull distribution with the same shape parameter, but a location parameter which will shift for different specimen volumes according to eq (4.31).

The integration in eq (4.28) has been performed by Weil and Daniel[8] for both the three-point and four-point flexure specimens with the result

$$B_{3f} = \frac{V_{3f}}{2(\alpha+1)^2}\left(\frac{\sigma_{3f}}{\hat{\sigma}}\right)^{\alpha} \tag{4.32}$$

for three-point flexure and

$$B_{4f} = \frac{V_{4f}(\alpha+2)}{4(\alpha+1)^2}\left(\frac{\sigma_{4f}}{\hat{\sigma}}\right)^{\alpha} \tag{4.33}$$

for the four-point flexure test at quarter points, where the subscripts $3f$ and $4f$ denote 3-point flexure and 4-point flexure, respectively. The stresses σ_{3f} and σ_{4f} are the stresses at the outer surface of the respective flexure specimens.

Equations (4.32) and (4.33) in conjunction with eq (4.27) yield

$$P(\sigma_{3f}) = \exp\left[-\left(\frac{\sigma_{3f}}{\hat{\sigma}_{3f}}\right)^{\alpha}\right] \tag{4.34}$$

$$P(\sigma_{4f}) = \exp\left[-\left(\frac{\sigma_{4f}}{\hat{\sigma}_{4f}}\right)^{\alpha}\right] \tag{4.35}$$

where

$$\hat{\sigma}_{3f} = \hat{\sigma}\left[\frac{2(\alpha+1)^2}{V_{3f}}\right]^{1/\alpha} \tag{4.36}$$

$$\hat{\sigma}_{4f} = \hat{\sigma}\left[\frac{4(\alpha+1)^2}{(\alpha+2)\,V_{4f}}\right]^{1/\alpha} \tag{4.37}$$

Thus, both the three-point flexure and four-point flexure tests yield two-parameter Weibull distributions with the same shape factor as the tensile test, but with different characteristic strengths.

Using eqs (4.30), (4.36) and (4.37), one obtains the following ratios for characteristic flexure strength to tensile strength

$$\frac{\hat{\sigma}_{3f}}{\hat{\sigma}_t} = [2(\alpha+1)^2(\frac{V_t}{V_{3f}})]^{1/\alpha} \tag{4.38}$$

$$\frac{\hat{\sigma}_{4f}}{\hat{\sigma}_t} = [\frac{4(\alpha+1)^2}{(\alpha+2)}(\frac{V_t}{V_{4f}})]^{1/\alpha} \tag{4.39}$$

In order to illustrate the effect of non-uniform stress distribution, consider the case $V_t = V_{3f} = V_{4f}$. For values of $\alpha = 15$ and 25, eqs (4.38) and (4.39) yield

$$\frac{\hat{\sigma}_{3f}}{\hat{\sigma}_t} = \begin{cases} 1.52, & \alpha = 15 \\ 1.33, & \alpha = 25 \end{cases} \tag{4.40}$$

$$\frac{\hat{\sigma}_{4f}}{\hat{\sigma}_t} = \begin{cases} 1.31, & \alpha = 15 \\ 1.20, & \alpha = 25 \end{cases}$$

These values of α are typical of currently utilized composites such as glass/epoxy and graphite/epoxy. Thus, the flexure test can produce significantly higher tensile strengths than the tensile test, with the three-point loading producing the highest strength. This is due to the fact that the maximum stress is produced at the outer surface in the center of the beam, while the four-point loading produces the maximum stress at the outer surface throughout the center section. In particular, the smaller the volume under maximum stress, the higher the local strength.

4.3.4 Off-Axis Tensile Test

The tensile test method described in paragraph 4.3.1 is based on ASTM Standard D3039-76, which is applicable to composites with orthotropic inplane properties. It is often desirable, however, to measure unidirectional tensile properties at some angle relative to the principal material coordinate system. This is accomplished by utilizing an off-axis tensile test. The off-axis test specimen consists of a laminate with all layers at equal fiber orientation, θ, with respect to the longitudinal axis of the specimen. This test method is often used to measure off-axis tensile modulus for the purpose of checking the validity of compliance transformation equations. It also provides a method of applying biaxial load relative to the principal material coordinate system.

Using eq (2.54) for the case of uniaxial tension, σ_x, in conjunction with the engineering constants, eqs (2.47), (2.58) and (2.61), one obtains

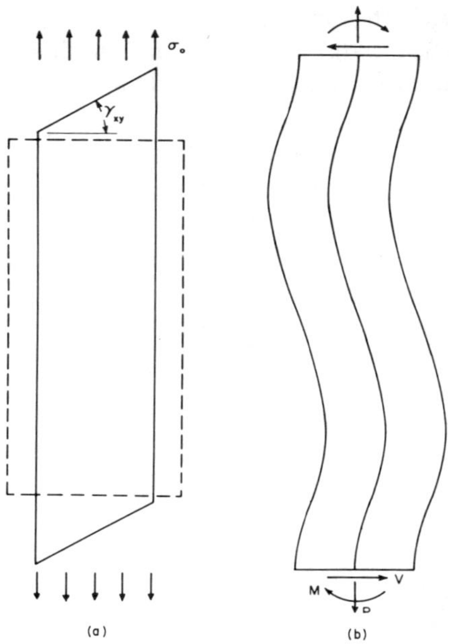

Fig. 4-13—Effect of end constraint on off-axis tensile specimen: (a) uniform state of stress, (b) effect of clamped ends (Pagano and Halpin[9])

$$\epsilon_x = \frac{1}{E_x}\sigma_x \tag{4.41}$$

$$\epsilon_y = \frac{-\nu_{xy}}{E_x}\sigma_x \tag{4.42}$$

$$\gamma_{xy} = \frac{\eta_{xy}}{E_x}\sigma_x \tag{4.43}$$

Thus, the material is anisotropic relative to the load direction and shear-coupling is present. It is the presence of the shear-coupling phenomenon that causes additional consideration in the off-axis tensile test compared to the tensile method for orthotropic composites covered by ASTM D3039-76. In particular, the clamping of the tensile coupon at the ends prohibits local rotation of the specimen which induces a non-uniform strain field as illustrated in Fig. 4-13. It has been shown by Pagano and Halpin,[9] however, that a uniform state of stress and strain will exist at the center of an off-axis tensile coupon if the length-to-width ratio, L/W, is sufficiently large. The exact dimensions necessary to accomplish this depend on the degree of shear coupling as measured by the value of η_{xy}. If an apparent modulus, E_x^* is experimentally determined in the usual manner for an off-axis tensile coupon, the error introduced by shear coupling can be estimated from the expression[9]

$$E_x = (1 - \eta)E_x^* \tag{4.44}$$

where

$$\eta = \frac{3\eta_{xy}}{\left(\frac{3E_x}{G_{xy}} + \frac{2L^2}{W^2}\right)} \tag{4.45}$$

Thus, η is a direct measure of the error involved in the observed modulus. A cursory examination of eq (4.45) reveals that η vanishes for decreasing values of η_{xy} or increasing ratios of L/W. The test procedure for the off-axis tensile test is essentially the same as described in D3039-76.

If the off-axis tensile test is used as a means of introducing biaxial stress relative to the principal material coordinate system, a three-element (rosette) gage is necessary to determine the stress–strain response parallel to the fibers, transverse to the fibers, and in longitudinal shear. The data reduction procedure consists of recording the load and the output of the three strain gages. Using eqs (2.34) for the case of uniaxial tension ($\sigma_y = \tau_{xy} = 0$) yields

$$\begin{aligned} \sigma_1 &= m^2\sigma_x \\ \sigma_2 &= n^2\sigma_x \\ \tau_{12} &= -mn\,\sigma_x \end{aligned} \tag{4.46}$$

If a rectangular rosette gage, as shown in Fig. 4-14, is used, the strain at 45 deg can be determined from eq (2.35) with the result

$$\epsilon_{45} = \frac{1}{2}\epsilon_x + \frac{1}{2}\epsilon_y + \frac{1}{2}\gamma_{xy} \tag{4.47}$$

where ϵ_{45} denotes the normal strain at $\theta = 45$ deg. Solving eq (4.47) for γ_{xy} yields

$$\gamma_{xy} = 2\epsilon_{45} - \epsilon_x - \epsilon_y \tag{4.48}$$

Substituting eq (4.48) into the transformation eq (2.35), one obtains

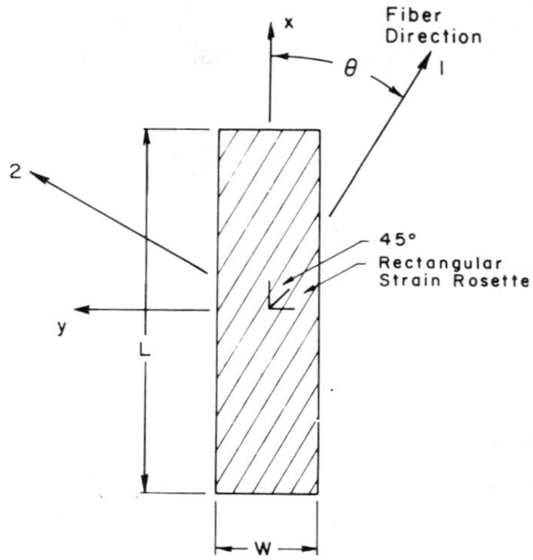

Fig. 4-14—Off-axis tensile coupon

$$\epsilon_1 = m(m - n)\epsilon_x + n(n - m)\epsilon_y + 2mn\epsilon_{45} \tag{4.49}$$

$$\epsilon_2 = n(m + n)\epsilon_x + n(n + m)\epsilon_y - 2mn\epsilon_{45} \tag{4.50}$$

$$\gamma_{12} = -(m^2 + 2mn - n^2)\epsilon_x - (m^2 - 2mn - n^2)\epsilon_y + 2(m^2 - n^2)\epsilon_{45} \tag{4.51}$$

Thus, the strains relative to the fiber orientation can be expressed in terms of the strain rosette measurements ϵ_x, ϵ_y, and ϵ_{45}. Equations (4.46) and (4.49)–(4.51) allow the stress–strain curves parallel to the fibers, transverse to the fibers, and in longitudinal shear to be determined under biaxial load. As will be discussed in a later paragraph, the off-axis tensile coupon can also be used as a means of measuring unidirectional shear response.

It should be noted that other three-element strain gage configurations can be used. The rectangular configuration, however, provides the simplest data reduction calculations and is commercially available.

In summary, the off-axis tensile test is run essentially the same as the tensile test for orthotropic composites as described in paragraph 4.3.1 with the exception of choice of specimen dimensions which must be designed to minimize shear coupling effects, and the use of a three-element strain gage if complete stress–strain behavior relative to the fiber orientation is desired. Equations (4.45) and (4.46) can be utilized in determining specimen dimensions. A rectangular rosette gage in

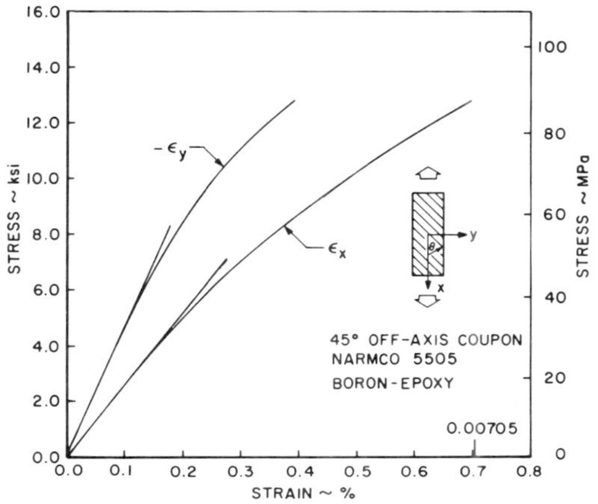

Fig. 4-15—Stress-stress response of 45-deg off-axis boron/epoxy tensile coupon (Pipes, Kaminski, and Pagano[62])

conjunction with eqs (4.49)-(4.51) provides a straight-forward means of determining strain response relative to the fiber orientation.

Typical off-axis stress-strain response is shown in Fig. 4-15 for a 45-deg boron-epoxy composite. A two-element rectangular strain gage was used to obtain response in the x and y directions.

4.3.5 Compression Test Methods

Perhaps the most difficult of the intrinsic material properties of composites to measure are the compressive strength properties. This is the case due to the fact that slight specimen geometric variations result in eccentricity of the applied load, thereby enhancing the opportunity for failure to occur due to geometric instability. Thus, in order to achieve an accurate measure of the compressive strength of a given composite material, rather complex loading fixtures and specimen configurations have been developed.

In this section three generic compression test methods will be discussed. The first method (Type I) is characterized by a specimen geometry having a test section length that is relatively short and is completely unsupported. The test specimen is generally loaded through friction by means of wedge action friction grips. Associated test fixtures, which insure colinearity of the applied load and specimen centerline, are required. To meet this need, several test-fixture configurations have been developed. One such fixture shown in Fig. 4-16 is the Celanese test fixture (ASTM D-3410-75). The Celanese fixture employs truncated

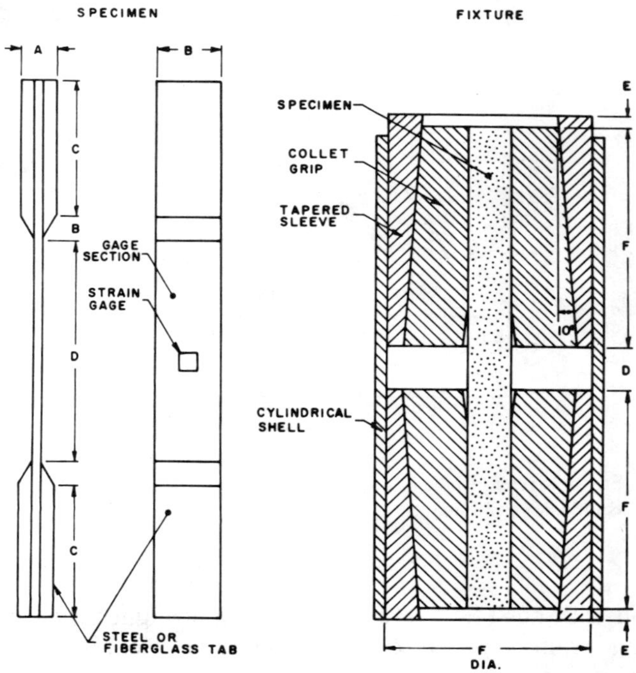

Dimension	mm	inches
A	39.9	0.157
B	6.35	0.250
C	57.2	2.250

Dimension	mm	inches
D	12.7	0.500
E	3.18	0.125
F	63.5	2.500

Fig. 4-16—Celanese compression specimen and test fixture, ASTM D 3410-75

conical friction grips contained in matching cylindrical end fittings. Colinearity of the cylindrical end fittings is insured by a hollow cylinder which contains the fittings. The test specimen geometry is also illustrated in Fig. 4-16 where the specimen is shown to be 141 mm (5.5 in.) in length and 6.4 mm (0.25 in.) in width with a test section length of 12.8 mm (0.5 in.). Beveled end tabs are bonded to the specimen for load introduction. Strain gages can be used with this specimen to measure modulus.

A second example of the Type I compression test method is the IITRI test method developed by Illinois Institute of Technology Research Institute.[10] The IITRI test fixture employs a test sample identical in geometry to that of the Celanese test method. Shown in Fig. 4-17, the IITRI test method employs linear bearings and hardened shafts to insure colinearity of the load path. Load is applied to the specimen

Fig. 4-17—IITRI compression test fixture (Hofer, Rao and Larsen[10])

through serrated wedges which are contained in solid steel bases. The IITRI fixture may also be modified to accept specimens of widths greater than 0.64 cm (0.25 in.). In addition, support rollers to restrain column instability of the specimen may be added to the wedge-action friction grips. These grip modifications are shown in Fig. 4-18.

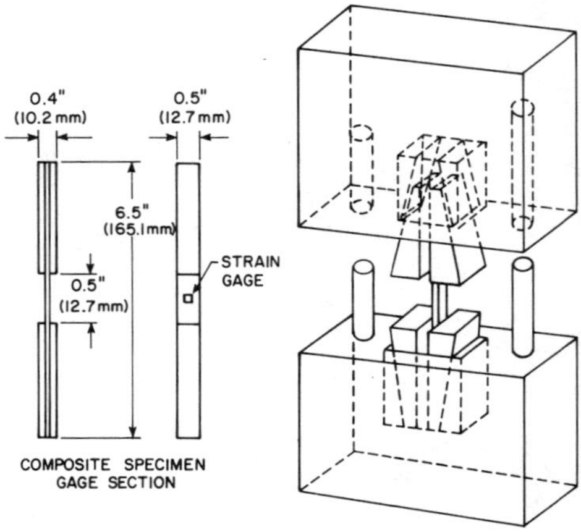

Fig. 4-18—Modified grips utilized in the IITRI compression test

Another example of the Type I compression test method is the Northrop method.[11] The Northrop compression test utilizes off-set unsupported lengths, as shown in Fig. 4-19. Stability is provided by the thick side supports. This method offers the advantage of a much simpler test fixture compared to the Celanese and IITRI methods.

The final example of the Type I compression test method was developed by the National Bureau of Standards (NBS).[12] The NBS test fixture combines certain features of the IITRI and Celanese test fixtures while introducing a feature which allows tensile loading. As shown in Fig. 4-20, the NBS compression test setup consists of a test specimen contained in end fixtures which are constrained to move in a colinear fashion by rigid rods and an external housing. Specimen gripping is achieved by friction due to interference between end fixtures and cylindrical epoxy specimen buildup. Overall specimen length is 118 mm (4.65 in.) while the test section is 16 mm (0.63 in.). This method utilizes both square cross-section and round cross-section specimens. The round cross section is recommended for 0-deg unidirectional composites only.

The four Type I compression test methods appear to yield acceptable data. However, certain problems may be encountered in their implementation. The Celanese test method requires extreme precision in mounting the specimen in the fixture. The mass of the IITRI test fixture requires prolonged soak periods for elevated temperature tests. All of the Type I test fixtures require that test specimens be fabricated such that specimen edges and/or tab surfaces are very close to perfect parallelism. Since the tabs are generally adhesively bonded to the specimen surfaces, bonding fixtures are required to insure the desired

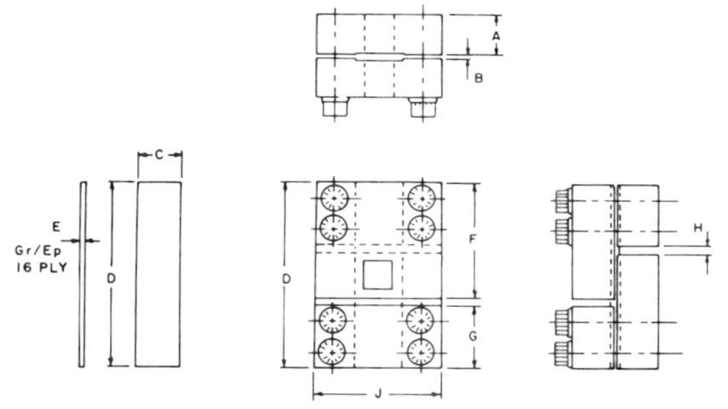

Dimension	Inches	mm
A	0.678	17.2
B	0.030	0.76
C	0.750	19.05
D	3.000	76.2
E	0.080-0.096	2.0-2.4
F	1.875	47.6
G	1.000	25.4
H	0.125	3.18
J	2.000	50.8

Fig. 4-19—Northrop compression specimen and test fixture (Verette and Labor[11])

parallelism. In addition, composite specimens fabricated with vacuum bag techniques may exhibit rather large thickness variations. Compressive tests of specimens prepared from these laminates may lead to erroneous results.

The second class of compression test methods (Type II) is characterized by a specimen of relatively long test section which is fully supported. The first example of the Type II compression test is the

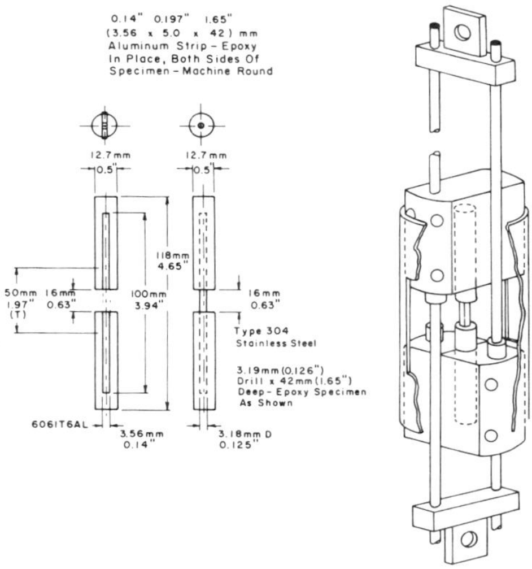

Fig. 4-20—NBS compression test fixture (Kasen and Schramm[12])

SWRI method developed by Southwest Research Institute.[13] The test specimen and support fixture are illustrated in Fig. 4-21. The SWRI test specimen geometry is a modification of the laminate tensile test specimen discussed earlier in which the load introduction tabs have been lengthened from 38 mm (1.5 in.) to 64 mm (2.5 in.) and the specimen length has been reduced from 230 mm (9 in.) to 218 mm (8 in.). The test fixture provides contact support for the specimen over the entire test section length. A notch is cut in one support to allow a transverse strain gage to be placed on the specimen for the purpose of measuring Poisson's ratio in compression. Longitudinal strain is determined by using a strain gage or extensometer placed on the edge of the specimen. For laminates displaying a significant free-edge effect, such longitudinal strain measurements may be erroneous.

A second example of the Type II compression test method is the side-supported fixture developed by Lockheed-California Company.[14] The specimen and fixture associated with this method are shown in Fig. 4-22. The Lockheed method utilizes side supports over the gage section of the specimen only. This is the major difference between this method and the SWRI method. In the Lockheed method the bottom end tab is cut off and the upper end tab, which extends outside the supports is gripped. Thus, the load is transferred through shear rather than direct compression. It should be noted that the bottom end tab does not have to be cut off. The specimen would then be gripped at both top and bottom.

It is often desirable to determine the compression strength utilizing a tensile coupon. Such is the case when residual compression strength is to be determined after pre-loading in tension[13] or after subjecting a tensile coupon to fatigue loading.[15] These requirements are a major reason for developing the Type II compression methods.

The Type II compression tests appear to yield laminate data which are compatible with Type I compression data. The values on 0-deg unidirectional compression strength generated with fully supported coupons, however, appear to be consistently lower than the Type I compression data. There is no precise explanation for this behavior. Perhaps the increased stiffness of the 0-deg unidirectional specimen makes alignment more critical in the case of the Type II compression method.

The third class of compression test method (Type III) involves the loading of a straight-sided coupon bonded to a honeycomb core, which supplies the support necessary for stability. The first example of the Type III compression test is the sandwich edgewise compression test which utilizes two coupons bonded to a honeycomb core, as shown in Fig. 4-23. Load is usually applied through self-aligning bearing blocks.

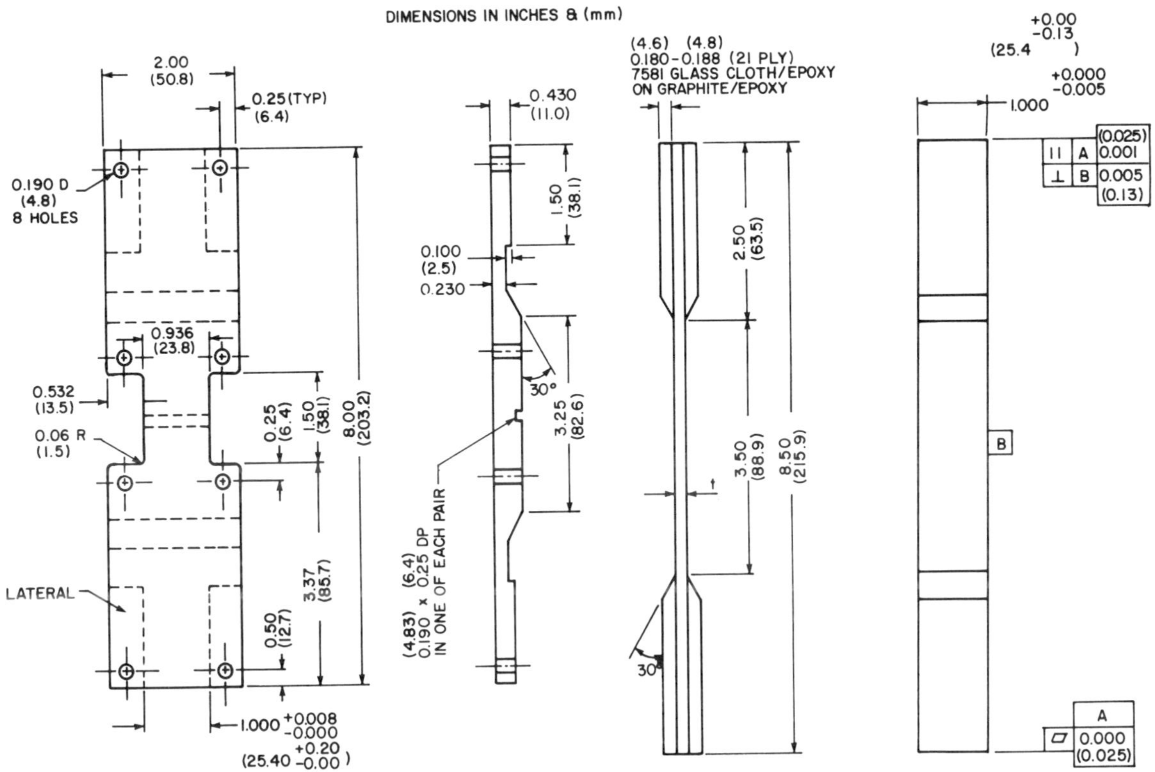

Fig. 4-21—SWRI fully supported compression specimen and fixture (Grimes et al[13])

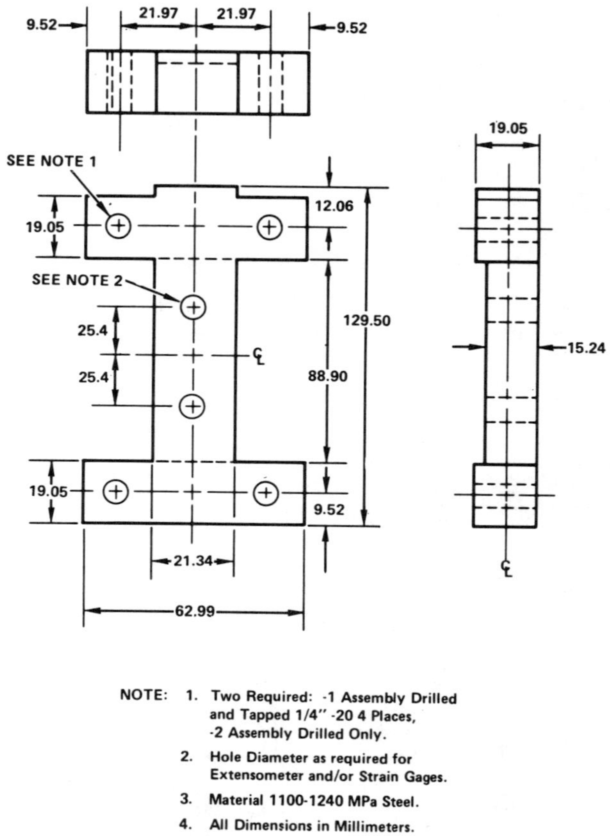

Fig. 4-22—Lockheed fully supported compression specimen and fixture (Ryder and Black[14])

A potting compound can be used to reinforce the ends, and straps can be clamped along the bottom of the flat laminate to provide a firm support for the loading heads and to prevent premature end crushing.[16] Stresses are calculated assuming that the core does not carry any load. Thus,

$$\bar{\sigma}_y = \frac{N_y}{h} = \frac{P}{2bh} \tag{4.52}$$

where $\bar{\sigma}_y$ is the average axial stress in the laminate. Strain gages mounted at the center of the face sheets can be used to measure E_y and ν_{yx} with the result

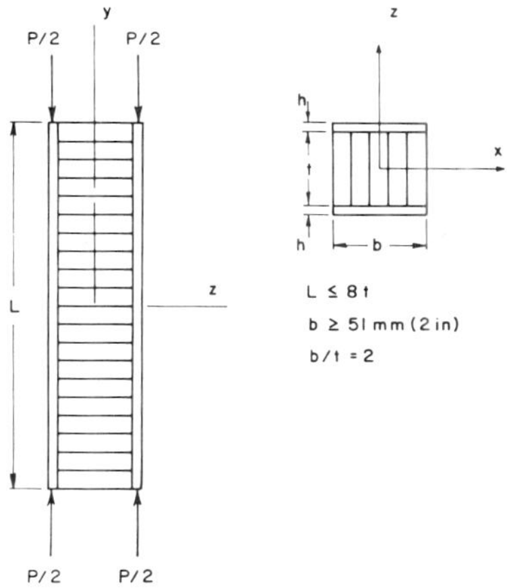

Fig. 4-23—Sandwich edgewise compression test specimen

$$E_y = \frac{P}{2bh\epsilon_y} \tag{4.53}$$

$$\nu_{yx} = \frac{-\epsilon_x}{\epsilon_y} \tag{4.54}$$

where ϵ_x and ϵ_y are the strains measured by the strain gages. The initial linear portion of the load-deformation curves are used to determine E_y and ν_{yx}. If the specimen is properly aligned, the opposite face strains should not vary more than 10 percent.[14]

The second example of the Type III test method involves the loading of a honeycomb sandwich beam in four-point bending with the face sheet on the compression side of the beam composed of the composite test specimen. A metal face sheet is used on the tension side of the beam. Core and face sheet materials, and beam dimensions are chosen such that failure occurs in the composite face sheet. Design parameters include metal face sheet strength; core shear strength, weight and cell size; strength of the adhesive and cure temperature; beam span and allowable mid-span deflection; and beam overhang and test section length. Some of these parameters depend on the specific composite material and laminate orientation, as discussed by Lantz.[16] The overall beam dimensions are usually constant, however, with the total length, including overhang, being 563 mm (22 in.), a 38-mm (1.5-in.) honeycomb core depth, and a face sheet width of 25.6 mm (1 in.). Further guidelines concerning core weight, beam overhang length, and beam span can be found in the Air Force Design Guide.[17]

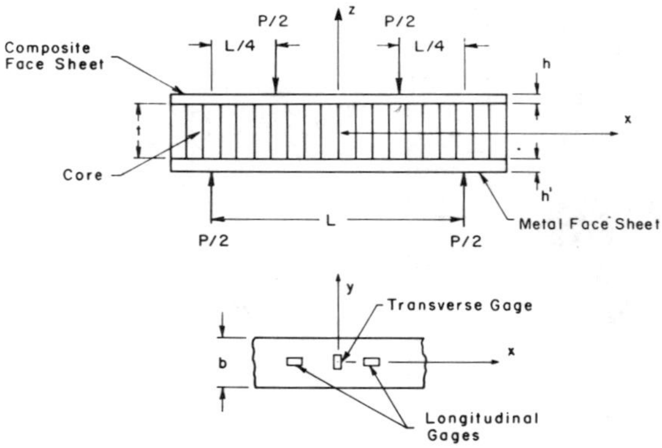

Fig. 4-24—Sandwich beam compression test specimen

As illustrated in Fig. 4-24, strain gages can be utilized at the center of the specimen to measure E_x and ν_{xy}. Data obtained from sandwich beam specimens for determination of ν_{xy} have shown values which are often higher than measurements obtained from tensile coupons. Boundary effects or the presence of transverse curvature are possible sources of Poisson's ratio error in the sandwich beam specimen.

Two methods have been utilized in transferring load in the sandwich beam test. The first method involves direct load application to the composite face sheet in a manner similar to the four-point flex test discussed in paragraph 4.3.2. If this method is used, hard rubber pads should be placed under the load noses to prevent local crushing of the composite face sheet. The second method utilizes load blocks buried in the core.[16] This eliminates direct bearing loads on the composite face sheet.

Laminate stress is determined by assuming uniform deformation in the face sheets and neglecting bending stresses in the core, with the result

$$\bar{\sigma}_x = \frac{N_x}{h} = \frac{PL}{4bh(2t + h + h')} \tag{4.55}$$

where $\bar{\sigma}_x$ is the average stress. The elastic constants are determined from the relationships

$$E_x = \frac{PL}{4bh\epsilon_x(2t + h + h')} \tag{4.56}$$

$$\nu_{xy} = \frac{-\epsilon_y}{\epsilon_x} \tag{4.57}$$

where ϵ_x and ϵ_y are the strains as measured by the strain gages at the center section of the composite face sheet. Again the initial linear portion of the load-deformation curves are used to determine E_x and ν_{xy}.

It should be noted that the sandwich beam can also be used to determine tensile properties.[18] Due to the success of the tensile coupon and the expensive nature of the sandwich beam specimen, the method is not widely utilized for tensile properties.

The sandwich beam specimen usually yields higher compression strength than any of the other methods.

4.3.6 Shear Test Methods

Shear tests are performed in order to establish the ultimate shear strengths, ultimate shear strains, and shear moduli of the composite material system. Proper characterization of the shear properties of the composite lamina or laminates requires evaluation of three distinct moduli and strengths. The lamina properties in the plane of lamination (1-2) are known as the inplane shear properties, while the properties in the 1-3 and 2-3 plane are known as interlaminar shear properties. To date, test methods have been developed for evaluation of the lamina inplane properties and for the lamina interlaminar properties in the 1-3 plane. No successful evaluation of the lamina interlaminar shear properties in the 2-3 plane has been accomplished to date. Shear tests for the composite laminate have generally been restricted to evaluation of properties in the plane of the laminate. No test method has been developed to date which yields intrinsic interlaminar properties for composite laminates. There are four generally accepted test methods for evaluation of lamina inplane shear properties. They are the $[\pm 45]_s$ coupon test, the off-axis coupon, the rail shear test method, and the torsion test. The $[\pm 45]_s$ coupon test method consists of the tensile test of a $[45/-45/45/-45]_s$ laminate having the following geometry:

overall length:	229 mm (9 in.)
test section length:	152 mm (6 in.)
width:	25 mm (1 in.)
top length:	38 mm (1.5 in.)

Note that this ± 45-deg coupon specimen geometry is identical to that given for the 90-deg tensile test in Section 4.3.1.

In order to explain how the tensile test of the $[\pm 45]_s$ laminate can be used to evaluate the lamina inplane shear properties, it will be necessary to review Sections 2.5.1 and 2.5.2 of Chapter 2. The state of strain for the $[\pm 45]_s$ laminate subjected to a uniaxial tensile stress, $\bar{\sigma}_x = N_x/h$, may be given as follows:

$$\begin{aligned} \epsilon_x^o &= H_{11} h \bar{\sigma}_x & \kappa_x &= 0 \\ \epsilon_y^o &= H_{12} h \bar{\sigma}_x & \kappa_y &= 0 \\ \gamma_{xy}^o &= 0 & \kappa_{xy} &= 0 \end{aligned} \tag{4.58}$$

Therefore, the state of stress within each of the lamina of the laminate may be written in terms of these strains.

$$\begin{aligned} \sigma_x &= \overline{Q}_{11}\epsilon_x + \overline{Q}_{12}\epsilon_y = h\bar{\sigma}_x(H_{11}\overline{Q}_{11} + H_{12}\overline{Q}_{12}) \\ \sigma_y &= \overline{Q}_{12}\epsilon_x + \overline{Q}_{22}\epsilon_y = h\bar{\sigma}_x(H_{11}\overline{Q}_{12} + H_{12}\overline{Q}_{22}) \\ \tau_{xy} &= \overline{Q}_{16}\epsilon_x + \overline{Q}_{26}\epsilon_y = h\bar{\sigma}_x(H_{11}\overline{Q}_{16} + H_{12}\overline{Q}_{26}) \end{aligned} \tag{4.59}$$

Note that since the shear coupling stiffness terms are equal and opposite for the +45-deg and −45-deg layers, i.e.,

$$\begin{aligned} \overline{Q}_{16}(45) &= -\overline{Q}_{16}(-45) \\ \overline{Q}_{26}(45) &= -\overline{Q}_{26}(-45) \end{aligned} \tag{4.60}$$

the shear stresses, τ_{xy}, in the +45 deg and −45 deg are equal and of opposite sign. Further, it can be shown that for $\theta = \pm 45$ deg, the normal stress σ_y vanishes and σ_x is equal to the applied stress, $\bar{\sigma}_x$

$$\begin{aligned} H_{11}Q_{12} + H_{12}Q_{22} &= 0 \\ H_{11}Q_{11} + H_{12}Q_{12} &= 1/h. \end{aligned} \tag{4.61}$$

Therefore, the state of stress within each of the layers is biaxial and consists of stress components σ_x and $\pm\tau_{xy}$. If the stresses are now transformed to the lamina coordinate system it is possible to determine the state of stress in the lamina coordinate system where the shear response is uncoupled from the normal response.

$$\begin{aligned} \sigma_1 &= m^2\sigma_x + 2mn\tau_{xy} \\ \sigma_2 &= n^2\sigma_x - 2mn\tau_{xy} \\ \tau_{12} &= -mn\sigma_x - (m^2 - n^2)\tau_{xy}. \end{aligned} \tag{4.62}$$

Now for $\theta = 45$ deg, $mn = n^2 = m^2 = 0.5$; therefore, eq (4.61) reduces to the following

$$\sigma_1 = (\sigma_x + \tau_{xy})/2$$

$$\sigma_2 = (\sigma_x - \tau_{xy})/2 \qquad (4.63)$$

$$\tau_{12} = \pm\,\sigma_x/2 = \pm\,\bar{\sigma}_x/2$$

The important result shown in eqs (4.63) is that while the normal σ_1 and σ_2 depend on both the applied stress, $\bar{\sigma}_x$ and the induced shear stress, τ_{xy}, the shear stress, τ_{12}, is a simple function of the applied stress, $\bar{\sigma}_x$. This is important because the induced shear stress, τ_{xy}, is statically indeterminant; that is, for nonlinear shear response typical of many composite systems, the magnitude of the induced shear stress cannot be determined. Thus, in regions of nonlinear material response it is not possible to determine the magnitude of either σ_1 or σ_2 as a simple function of the applied stress, $\bar{\sigma}_x$. Note that for any angle, θ, other than ± 45 deg, the inplane shear stress, τ_{12}, is also indeterminant.

Consider now transformation of the laminate state of strain (ϵ_x^o, ϵ_y^o) to the lamina coordinate system.

$$\epsilon_1 = (\epsilon_x^o + \epsilon_y^o)/2$$

$$\epsilon_2 = (\epsilon_x^o + \epsilon_y^o)/2 \qquad (4.64)$$

$$\gamma_{12} = -(\epsilon_x^o - \epsilon_y^o)$$

Therefore, by performing the tensile test of the $[+45/-45/45/-45]_s$ laminate and monitoring the applied stress $\bar{\sigma}_x$ and the laminate longitudinal and transverse strains, ϵ_x^o and ϵ_y^o, it is possible to establish the lamina inplane shear response.

$$\tau_{12} = \bar{\sigma}_x/2$$

$$\gamma_{12} = (\epsilon_x^o + \epsilon_y^o) \qquad (4.65)$$

Further, if the laminate effective modulus, E_x, and Poisson's ratio, ν_{xy}, are defined as follows:

$$E_x = E_{\pm 45} = \bar{\sigma}_x/\epsilon_x^o$$

$$\nu_{xy} = \nu_{\pm 45} = -\,\epsilon_y^o/\epsilon_x^o \qquad (4.66)$$

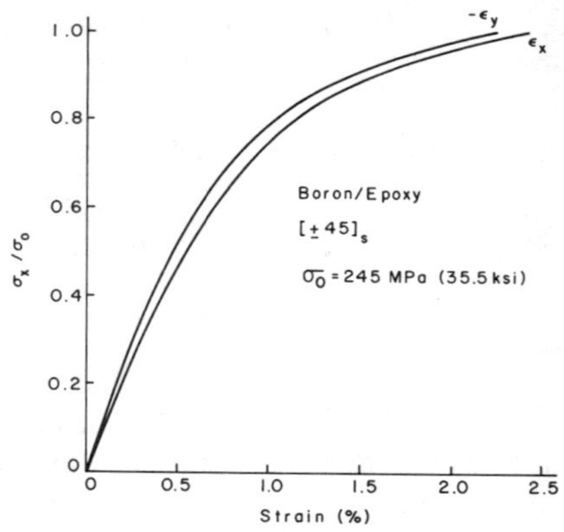

Fig. 4-25—Stress–strain response for $[\pm 45]_s$ boron/epoxy laminate (data from Rosen[19])

an expression for the inplane shear modulus, G_{12}, in terms of the $[\pm 45]_s$ laminate effective properties may be developed.

$$G_{12} = \frac{E_x}{2(1 + \nu_{xy})} \tag{4.67}$$

Typical stress–strain results for the $[+45/-45/45/-45]_s$ laminate tensile test of a boron-epoxy material system are shown in Fig. 4-25. Transformation of these results to lamina inplane shear response by eq (4.65)

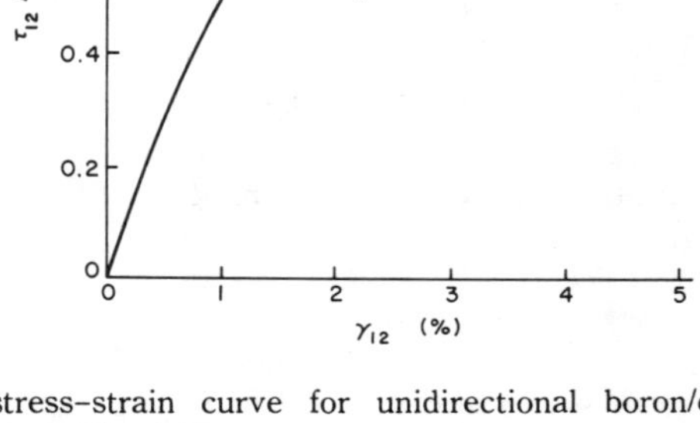

Fig. 4-26—Shear stress–strain curve for unidirectional boron/epoxy composite as derived from Fig. 4-25

is shown in Fig. 4-26. This test procedure was suggested by Rosen[19] and details are presented in ASTM Standard D3518-76.

It should be noted that although the $[\pm 45]_s$ laminate tensile test can be employed to establish shear stress–strain response well into the region of nonlinear material response, caution must be exercised in interpretation of the ultimate stress and strain results. This is due to the fact that the lamina is in a state of combined stress rather than pure shear. Hence, it should be expected that the presence of the normal stress components would have a deleterious effect upon ultimate shear strength. In particular the transverse normal stress, σ_2, can be expected to significantly alter the apparent shear strength. Data which tend to confirm the expected influence of transverse normal stress upon shear strength were developed by comparing the ultimate strengths of the $[\pm 45]_s$ laminate in tension and compression. When the laminate is loaded in tension, the normal stress components are tensile, while a compression loading leads to compressive normal stresses. Since it is expected that the presence of a tensile transverse normal stress would reduce the apparent inplane shear strength, while a compressive normal stress may result in an increase in strength, a difference in compressive and tensile strengths of the $[\pm 45]_s$ laminate would support the argument of the deleterious influence of the tensile transverse stress upon ultimate shear strength. Such data are presented in Table 4-1 for a graphite-epoxy material system where it can be seen that the average tensile

Table 4-1 Shear Strength from $[\pm 45]_s$ Laminate Test

AS/3501 graphite/epoxy

	Tension MPa (ksi)	*Compression* MPa (ksi)
	172 (24.9)	203 (29.4)
	177 (25.6)	206 (29.9)
	163 (23.6)	193 (27.9)
	155 (22.5)*	204 (29.5)
	148 (21.5)	202 (29.3)
	144 (20.9)*	203 (29.4)
	152 (22.0)*	
	155 (22.5)*	
	148 (21.5)*	
avg.	157 (22.8)	201 (29.2)

**8 ply laminate*

strength for the $[\pm 45]_s$ laminate is 157 MPa (22.8 ksi), while the average compressive strength is 201 MPa (29.2 ksi).

The second method for determination of lamina shear response is the off-axis tensile test, which was discussed in paragraph 3.3.4. For shear response relative to the fiber direction under the uniaxial load σ_x, eqs (4.47) and (4.51) yield

$$\tau_{12} = -mn\sigma_x$$

$$\gamma_{12} = -(m^2 + 2mn - n^2)\epsilon_x - (m^2 - 2mn - n^2)\epsilon_y + 2(m^2 - n^2)\epsilon_{45} \tag{4.69}$$

As previously discussed, eq (4.69) is based on the use of a rectangular three-element rosette gage. If eqs (4.68) and (4.69) are specialized for $\theta = 45$ deg, then $m^2 = n^2 = mn = 0.5$ and

$$\tau_{12} = \frac{-\sigma_x}{2} \tag{4.70}$$

$$\gamma_{12} = \epsilon_y - \epsilon_x \tag{4.71}$$

Using eqs (4.41) and (4.42) in conjunction with eq (4.71) yields

$$\gamma_{12} = -(1 + \nu_{xy})\frac{\sigma_x}{E_x} \tag{4.72}$$

Combining eqs (4.70) and (4.72) allows determination of the lamina inplane shear modulus, G_{12} in terms of the 45-deg off-axis laminate properties.

$$G_{12} = \frac{\tau_{12}}{\gamma_{12}} = \frac{E_x}{2(1 + \nu_{xy})} \tag{4.73}$$

It should be noted that eq (4.73) is identical to that developed for the $[\pm 45]_s$ tensile test and given in eq (4.67). The response of the 45-deg off-axis and $[\pm 45]_s$ laminate test specimens is, however, quite different. For example, the effective properties of the two test specimens for a given boron-epoxy material are given as follows:

$$E_x(\pm 45) = 20.98 \text{ GPa } (3.04 \times 10^6 \text{ psi})$$
$$\nu_{xy}(\pm 45) = 0.69$$

$$G_{12} = \frac{(20.98)}{2(1 + 0.69)} = 6.21 \text{ GPa } (0.90 \times 10^6 \text{ psi})$$

$$E_x(45) = 15.87 \text{ GPa } (2.30 \times 10^6 \text{ psi})$$
$$\nu_{xy}(45) = 0.28$$

$$G_{12} = \frac{(15.87)}{2(1 + 0.28)} = 6.21 \text{ GPa } (0.90 \times 10^6 \text{ psi})$$

Note that both of the effective elastic properties for the $[\pm 45]_s$ laminate test are greater than the corresponding properties of the 45-deg off-axis laminate, yet the same value of G_{12} is determined by each of the test methods. As previously discussed, eq (4.45) can be utilized in designing the dimensions of the off-axis tensile specimen to assure that the shear coupling effects are minimized.

The inplane shear stress–strain response, as determined in the off-axis test, is consistent for any angle, θ, as shown in Fig. 4-27 for boron-epoxy. As noted in this figure, however, the ultimate shear strength is a function of the fiber orientation of the off-axis specimen. Although the analysis presented earlier indicates that the 45-deg off-axis specimen is attractive because the relation between the lamina shear response and the 45-deg off-axis lamina response is quite simple, the results presented in Fig. 4-27 show that ultimate shear strength predicted by the 45-deg specimen may be in error by as much as 30-40 percent. Hence, this lead Chamis and Sinclair[20] to recommend the 10-deg off-axis test for determining lamina shear properties. The 10-deg angle was chosen to minimize the effects of longitudinal and trans-

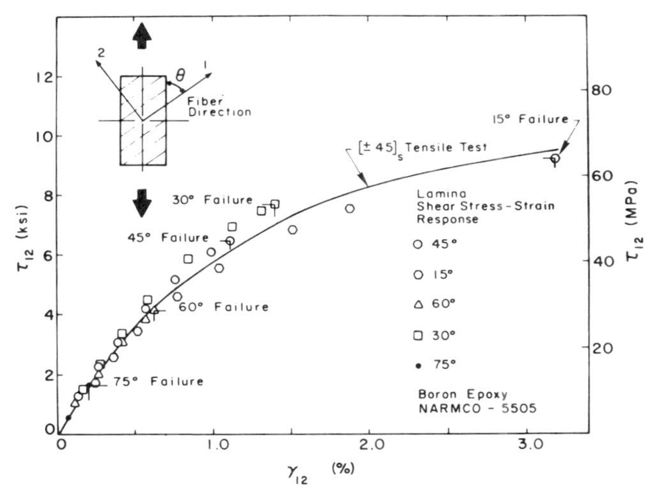

Fig. 4-27—Shear stress–strain curve generated from off-axis tensile test (Pipes and Cole[63])

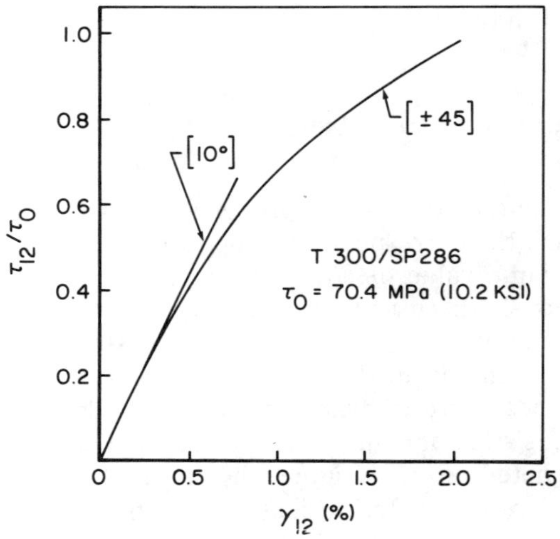

Fig. 4-28—Comparison between $[\pm 45]_s$ laminate 10 deg off-axis methods for determining shear stress–strain curve in graphite/epoxy (data from Daniel[21])

verse tension components σ_1 and σ_2 on the shear response. Results for the 15-deg off-axis test shown in Fig. 4-27 indicate that the ultimate shear strength determined by the 15-deg test and the $[\pm 45]_s$ tensile test compare quite favorably. A comparison between the 10-deg off-axis method and the $[\pm 45]_s$ laminate method for determining shear stress–strain behavior is shown in Fig. 4-28 for graphite-epoxy.[21] Note that the 10-deg off-axis test yields a higher initial modulus, while the $[\pm 45]_s$ laminate yields a higher strength. The $[\pm 45]_s$ laminate test yields considerably more of the shear stress–strain curve than the 10-deg off-axis test.

The third method for determination of lamina inplane shear response is the rail shear test method. Two test specimens and fixture configurations are generally accepted, the two-rail fixture shown in Fig. 4-29, and the three-rail fixture shown in Fig. 4-31. The test fixture consists primarily of rigid steel rails which are bolted and/or bonded to a rectangular laminate specimen. For determination of lamina shear properties the laminate consists of 8-12 laminae whose fiber orientation is parallel or perpendicular to the longitudinal axis of the rails. It should be noted, however, for the case where the fibers are perpendicular to the rails, the clamping of the fibers may induce a higher shear strength compared to the case where the fibers are parallel to the rails. A $[0/90]_s$ laminate may also be utilized for unidirectional characterization, since the shear response should be identical for 0- and 90-deg orientations

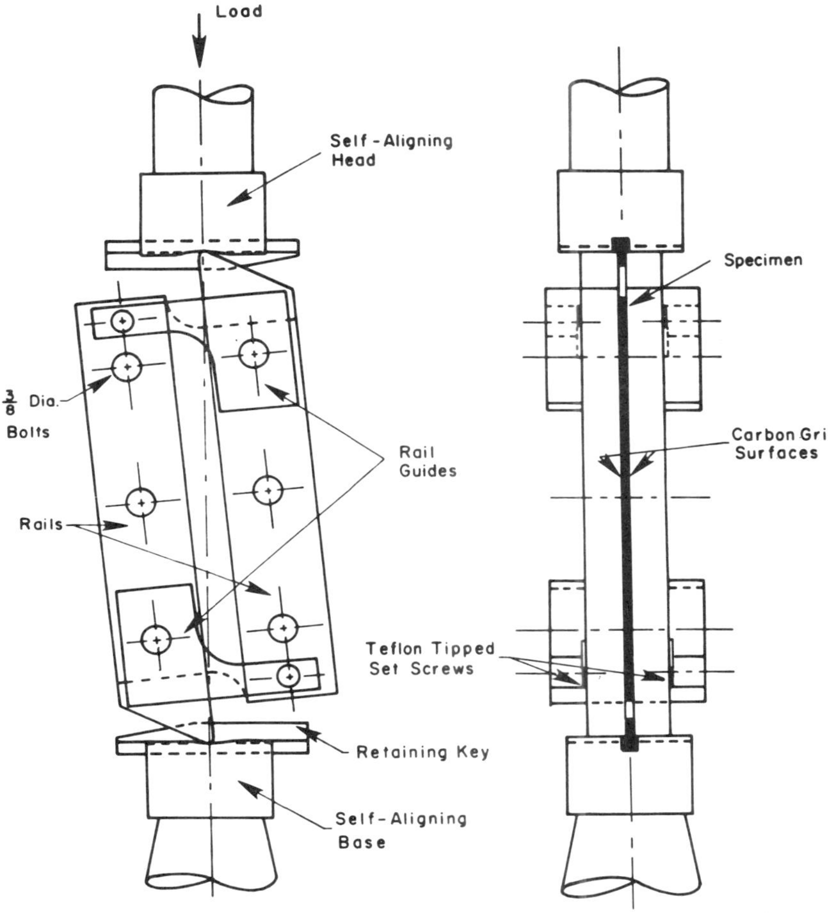

Fig. 4-29—Two-rail shear apparatus (ASTM D-30 committee, unpublished)

(e.g., $A_{66} = G_{12}h$ for $[0/90]_s$ laminates). Other laminates having orthotropic inplane properties ($A_{16} = A_{26} = 0$) relative to an x-y axis system parallel and perpendicular to the rails, respectively, may also be utilized in this method.

The test specimen geometry for the two-rail and three-rail configurations is shown in Fig. 4-30 and 4-32, respectively. The shear stress calculation for the two configurations may be expressed as a function of the applied load, P, the laminate thickness, h, and the width between the rails, b.

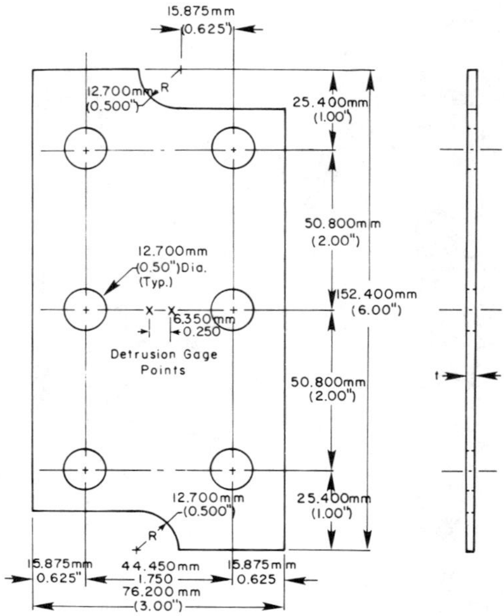

Fig. 4-30—Two-rail shear specimen (ASTM D-30 committee, unpublished)

$$\bar{\tau}_{xy} = \frac{N_{xy}}{h} = \frac{P}{bh} \quad \text{(two-rail)}$$
$$\bar{\tau}_{xy} = \frac{N_{xy}}{h} = \frac{P}{2bh} \quad \text{(three rail)} \tag{4.74}$$

Shear strain can be measured with a rectangular rosette gage located at the geometric center of the specimen. For the case of pure shear

$$\epsilon_x = \epsilon_y = 0 \tag{4.75}$$

and eq (2.38) yields

$$\gamma_{xy} = 2\epsilon_{45} \tag{4.76}$$

Thus, in theory, a single-element gage at 45 deg is sufficient to determine shear strain. The three-element gage, however, serves as a means of assuring that the strain field in the center of the specimen is pure shear (i.e., the longitudinal and transverse gages should show essentially zero strain). It should be noted that the free edges at the top and bottom of the rail shear specimen induce large normal stresses con-

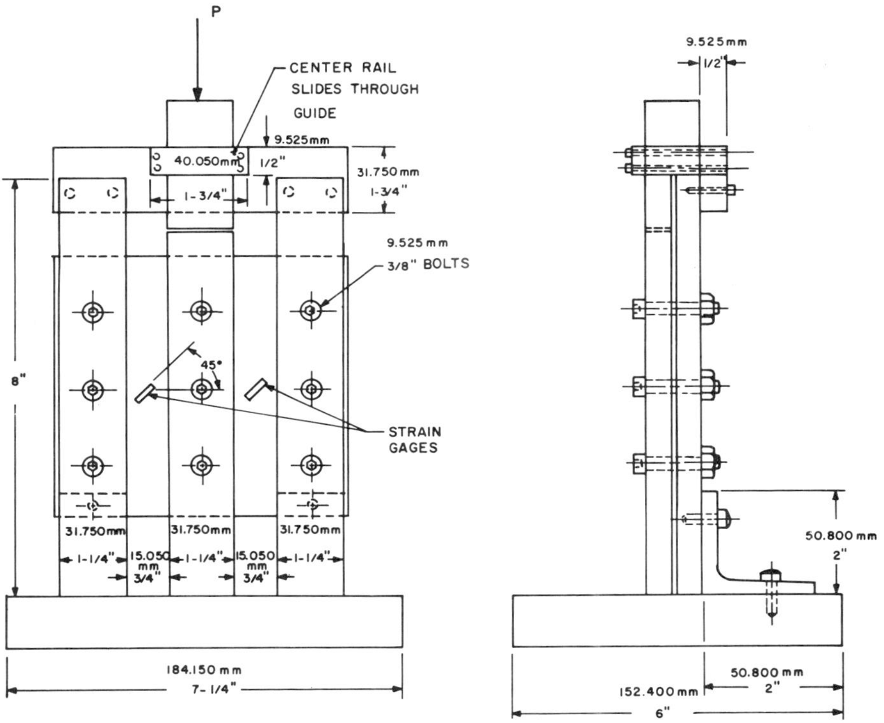

Fig. 4-31—Three-rail shear apparatus (ASTM D-30 committee, unpublished)

centrated in the corners.[22] If the length-to-width ratio of the gage section is sufficiently large, 8:1, a state of uniform shear will exist in the center of the gage section. These results are obtained on the basis that the edges are perfectly clamped. Thus, a means of checking the strain field in the center of the gage section is desirable. The perfectly clamped boundary assumption requires that the bolts in the rail shear specimens apply even clamping pressure to the edges. The laminate shear modulus, $G_{xy} = A_{66}/h$, can be determined from the slope of the initial straight-line portion of the shear stress–strain curve.

The last method for determining inplane shear properties of a laminate is the torsion method, which can be utilized in conjunction with either a solid rod (unidirectional specimen only) or a tubular specimen.

In the solid rod method[23] a unidirectional specimen is fabricated by either laying-up the material in a cylindrical mold or machining the specimen from a solid rectangular bar. This test is not widely used,

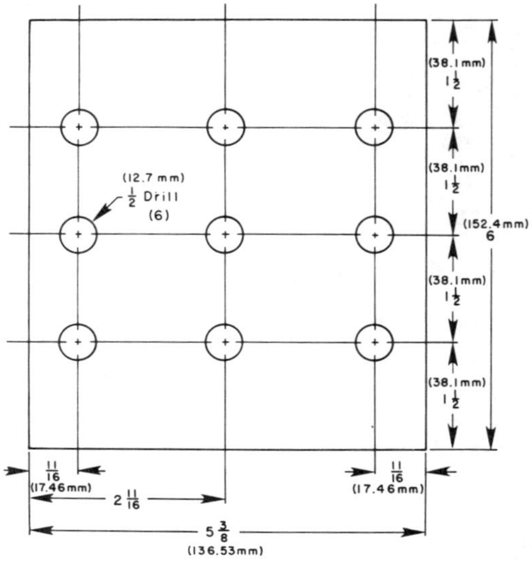

Fig. 4-32—Three-rail shear specimen (ASTM D-30 committee, unpublished)

however, because other methods have much more advantage. In particular, a solid rod under torsion yields a shear stress distribution which is linear with respect to the radial distance from the center of the rod. Thus, the calculation for stress is accurate only for linear portions of the load-deformation curve. With the large nonlinear response often observed for many unidirectional shear stress–strain curves, the usefulness of the test is highly reduced. In addition, the solid rod specimen is not easy to fabricate. Other approaches, such as the $[\pm 45]_s$ tensile test, do not have these drawbacks.

Torsion of a thin walled tube provides a means of directly applying pure shear to a fiber reinforced composite specimen. The major drawback to this method, however, is the cost and difficulty associated with fabricating quality tubular specimens. In addition, the method requires specialized equipment and gripping systems. Problems associated with end fixtures are discussed in greater detail in conjunction with a later section on combined loading of tubular specimens. For the case of pure torsion, the end attachment problem is much less severe than for the case of combined loading. A simple and effective grip design involves metal grips bonded to the tube ends with an epoxy potting material.[24,25] The grips are bonded on both the outside and inside of the tube to increase the area of load introduction. There are a number of design configurations that these grips can take. One of the most efficient, from both a load introduction and specimen alignment standpoint, is the grip design chosen by Hahn and Erikson,[25] as shown in Fig. 4-33. This grip consists of concentric cylinders with the tubular specimen bonded between the cylinders. Two pins through the outer cylinder are used in conjunction with the fixture. One pin connects the inner cylinder with the outer cylinder and the second connects the entire fixture and specimen to the loading machine. This type of fixture has the advantage of not requiring reinforced ends.

For the case of orthotropic laminates ($A_{16} = A_{26} = 0$) of symmetric construction ($B_{ij} = 0$) subjected to pure torsion[26]

$$\bar{\tau}_{xy} = \frac{N_{xy}}{h} = \frac{T}{2\pi R^2 h} \tag{4.77}$$

where T is the applied torque and R is the mean radius of the cylinder. The shear strain is related to the angle of twist per unit length, ϕ, by

$$\gamma^o_{xy} = R\phi \tag{4.78}$$

Thus, a shear stress–strain curve can be obtained from a plot of torque versus angle of twist. The angle of twist, however, contains effects of machine compliance and is not a good measure of shear strain. Much more accurate results can be obtained by using a strain gage. As in the case of rail shear, a rectangular rosette gage provides a direct measure-

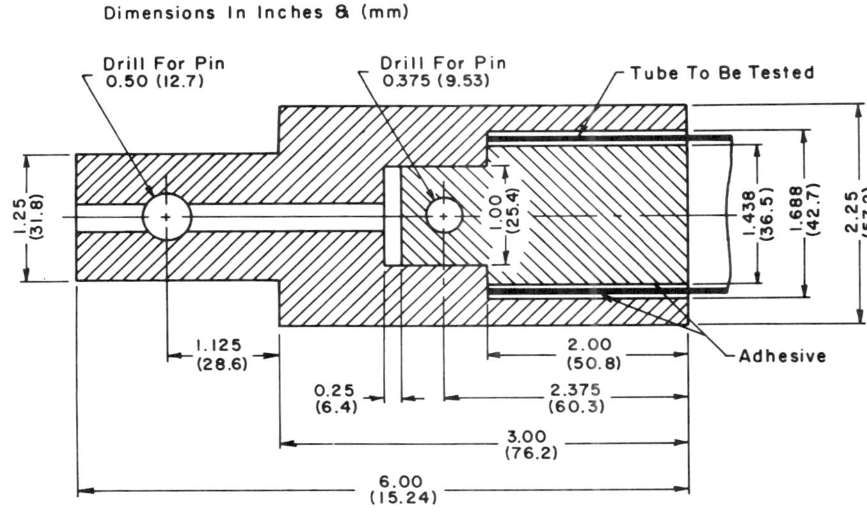

Fig. 4-33—Tubular end attachment (Hahn and Erikson[25])

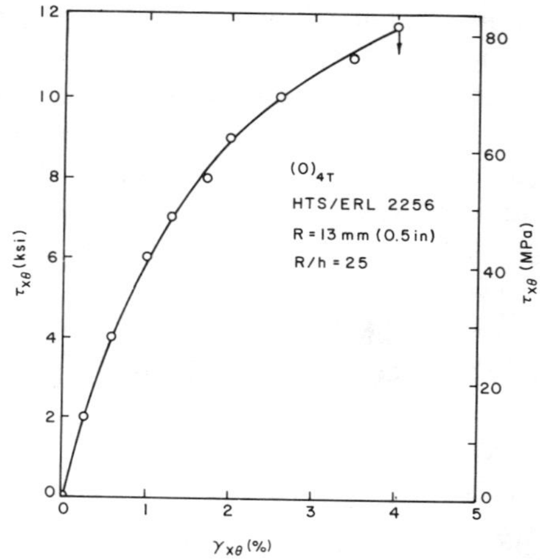

Fig. 4-34—Shear stress–strain curve for graphite-epoxy unidirectional tube (Grimes and Francis[30])

ment of shear strain, i.e., eq (4-76) can be used to determine the shear strain in the tube. Again the longitudinal and transverse gages should register zero for pure shear. It should also be noted that a strain gradient exists in a composite tubular specimen.[27,28] This gradient can be approximated as a linear function of the z coordinate located in the center of the tube wall and directed normal to the outer surface of the cylinder.[29]

$$\gamma_{xy} = \left(\frac{1+z}{R}\right) \gamma^o_{xy} \tag{4.79}$$

This gradient can be checked experimentally by using gages both inside and outside of the tube. If the shear strain is measured on the outer surface, then eq (4.79) yields

$$\gamma^o_{xy} = \frac{\gamma_{xy}}{(1 + 2h/R)} \tag{4.80}$$

In order to minimize the effect of the strain gradient on the ply stresses, $R/h \geq 10$. In addition to insure sufficient gage section in the tube, $L/R \geq 8$, where L denotes the length of the tube between end fixtures.

The composite shear modulus, $G_{xy} = A_{66}/h$, can be obtained by measuring the slope of the initial straight-line portion of the $\bar{\tau}_{xy} - \gamma^o_{xy}$ stress–strain curve.

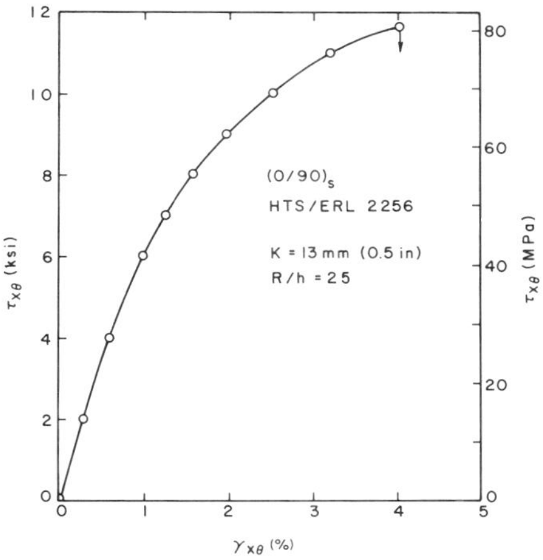

Fig. 4-35—Shear stress–strain curve for graphite/epoxy bidirectional laminate (Grimes and Francis[30])

Shear stress–strain curves are shown in Figs. 4-34 and 4-35 for a $[0]_{4T}$ unidirectional tube and a $[0/90]_s$ laminated tube fabricated from Hercules HTS graphite fiber in Union Carbide's ERL 2256 epoxy resin.[30] As predicted from laminated plate theory, both of the shear stress–strain curves are almost identical.

4.3.7 Interlaminar Beam Tests

Unlike the shear test methods discussed in the previous section, interlaminar beam tests are utilized to estimate interlaminar shear strength only. These methods do not provide any measurement of shear modulus.

The short beam shear test has become a widely used method for characterizing the interlaminar failure resistance of fiber-reinforced composites. This test method involves a three-point flexure specimen with the span-to-depth ratio, L/h, chosen to produce interlaminar shear failure. This method is subject to the same restrictions with regard to material applicability as discussed in Section 4.3.2 for the general flexure test. An additional complexity is presented by the short beam shear method when used in conjunction with laminated materials. In particular, the interlaminar shear stress will be parabolic within each layer, but a discontinuity in slope will occur at the ply interface. As a result the maximum shear stress will not necessarily occur at the center of the beam, as shown for the $[0/90]_s$ graphite-epoxy laminate in Fig. 4-36.[3]

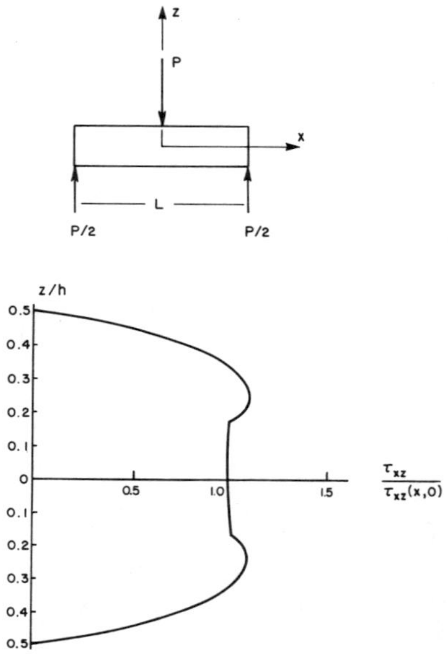

Fig. 4-36—Interlaminar shear stress distribution for $[0/90]_s$ graphite/epoxy laminate

As in the case of in-plane ply stresses, laminated beam theory is needed to calculate the stresses. Thus, the short beam method is applicable only to composite materials which can be treated as homogeneous.

The simplicity of the short beam shear method makes it a very popular materials screening tool. As pointed out in the title of ASTM Standard D-2344, this method measures the "apparent" interlaminar shear strength of composite materials. Thus, the short beam shear method is not appropriate for generating design information. Despite such warnings data generated from this method have had a tendency to be used as design allowables.

A second limitation on this test method, in conjunction with graphite-epoxy materials, raises serious doubts about its usefulness even as a materials screening tool. In particular, when used in conjunction with thin unidirectional beams (16 plies), which is common practice with high-performance graphite-epoxy composites, the test method does not yield interlaminar failures. Such data are often reported in the literature without mentioning the failure mode, leaving the reader to believe that the desired interlaminar failure was attained. Photomicrographs of typical graphite-epoxy 16-ply short beam shear failures show a compressive buckling failure near the load nose.[64]

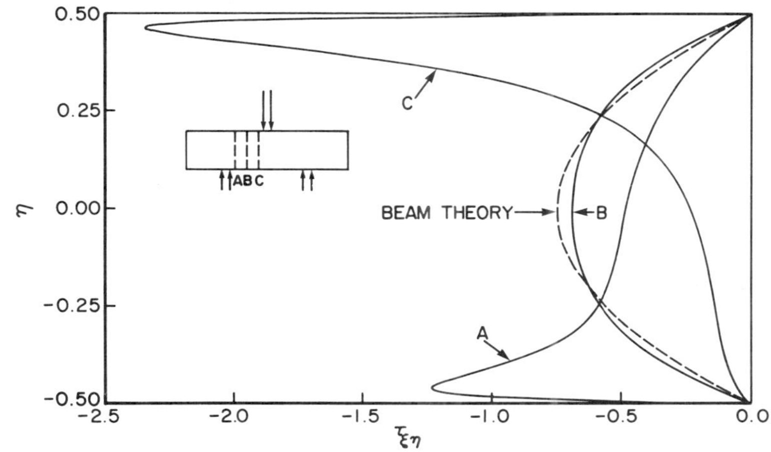

Fig. 4-37—Shear stress distribution in 50-ply graphite/epoxy beam, $L/h = 4$

Precise dimensions recommended for the short beam shear test can be found in ASTM D-2344. A specimen width and maximum thickness of 64 mm (0.25 in.) is required by the ASTM standard. There is no minimum thickness required, and as a result short beam shear specimens as thin as 2 mm (0.08 in.) have been tested. This means at $L/h = 4$, the usual span-to-depth ratio for graphite-epoxy unidirectional composites, a load nose of 6.4-mm (0.25-in.) diameter will cover the entire span length creating a "punchout" test rather than a flexure test. Thicker specimens (50 plies) utilized in conjunction with graphite-epoxy unidirectional composites have revealed interlaminar failures.[65] These interlaminar cracks were preceded, however, by vertical fracture near the load nose.[64]

In order to avoid the difficulty associated with small specimens, including measurement of interlaminar shear stiffness, a very large interlaminar shear beam was investigated by Pipes, Reed, and Ashton.[31] These beams were 13-mm (0.5-in.) wide, 25-mm (1-in.) thick, and 102-mm (4-in.) long. Although this approach alleviates the problem associated with excessively small values of L resulting from low span-to-depth ratios, the problem of large shear stresses in excess of those predicted by classical beam theory is present. A finite-element analysis performed by Berg, Tirosh, and Israeli,[32] shows such large shear stresses in the vicinity of the applied center load. Similar results were obtained by Pipes, Reed, and Ashton.[31] A recent Fourier series solution to short beam shear experiments,[66] reveals that classical beam theory is never completely recovered anywhere along the span of the beam. Typical results are shown in Fig. 4-37 for a 50-ply graphite-epoxy beam. Note that the shear stress distribution becomes skewed near the load nose and supports. The symbol η denotes a normalized through-the-thickness coordinate, while $\tau_{\xi\eta}$ is the interlaminar shear stress normalized by the maximum value of the corresponding stress as determined from classical beam theory.

An alternative to the classical short beam shear test has been suggested by Browning, Abrams, and Whitney.[65] The proposed method utilizes a four-point bend specimen with quarter-point loading and a span-to-depth ratio which is considerably less than utilized for flexural strength. For example, a span-to-depth ratio of 16 was utilized with graphite-epoxy unidirectional material rather than the standard value of 32 employed in conjunction with flexural strength determinations. Interlaminar fractures were observed. However, as in the case of the thick short beam shear failures, the horizontal split was usually preceded by a vertical crack near the load nose where the interlaminar failure was observed to initiate. This interlaminar test method has been referred to as the four-point shear test and has the advantage of producing interlaminar failures with a thin specimen (16 plies). This method will usually produce interlaminar failures for any material in which an interlaminar failure can be obtained in conjunction with a thick short beam shear specimen.

For both three-point bending and four-point bending, the apparent interlaminar shear strength is determined from the relationship

$$\tau_m = \frac{3P}{4bh} \tag{4.81}$$

where τ_m is the maximum value of τ_{xz}. Using eqs (4.25) and (4.26), it can easily be seen that classical beam theory predicts an interlaminar shear failure if

$$L/h < S_x/S_{xz} \tag{4.82}$$

where S_x and S_{xz} are the tensile strength (or compressive strength, whichever is the lesser) and interlaminar shear strength, respectively.

The difficulties associated with interlaminar beam tests have not influenced their popularity. Simplicity of testing and the inexpensive nature of the specimens make the methods an ideal tool for quality control or materials screening. Despite their wide use, especially the short beam method, the value of interlaminar beam tests remains highly questionable.

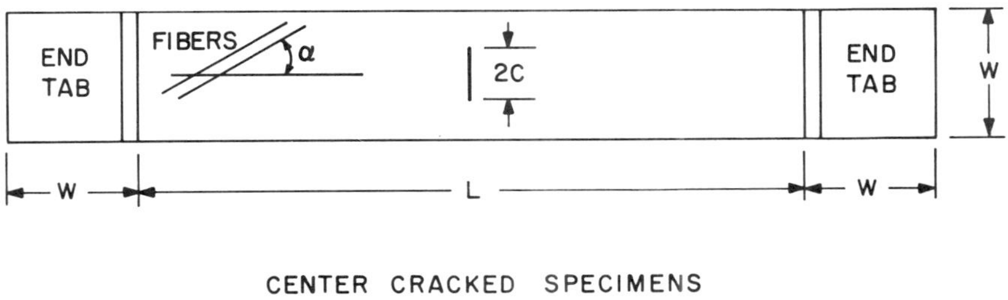

Fig. 4-38—Center cracked specimens (Nuismer and Whitney[35])

4.4 FAILURE CHARACTERIZATION OF COMPOSITES

In section 3.3, the more routine strength characterization test methods, including total stress–strain behavior, were discussed. Characterization of the failure behavior under conditions which are considered much less routine in nature is the subject matter of the present section. Notch strength characterization, determination of first ply failure, strength characterization under combined loading and determination of fatigue behavior all fall in the category of non-routine material property characterization.

4.4.1 Composite Notched Strength

Experimental data on notched composite laminates have been limited primarily to straight cracks and circular holes. Details of experimental procedures, including data reduction and comparison to theory, will be discussed for these two notch geometries.

Although a number of test specimen geometries, including the three-point bend test[33] and the side notch,[34] have been used for determining the strength reduction of a composite laminate containing a through-the-thickness crack, the center notch specimen is the most widely used. A straight-sided tensile coupon, as shown in Fig. 4-38, is utilized for this test. The cracks are typically formed by drilling a 0.01-inch-diameter pilot hole and then using 0.005-inch-diameter diamond wire to complete the crack.[35] The cracks can also be machined ultrasonically.[21]

The gross failure stress, σ_N (failure load divided by unnotched cross-sectional area), is obtained for notched specimens. These values can be adjusted to obtain σ_N^∞ from the relationship[36]

$$K_Q = Y_1(2c/W)\sigma_N\sqrt{\pi c} \tag{4.83}$$

where Y_1 is the finite-width correction factor for isotropic materials. Equation (4.83) in conjunction with eq (2.138) yields

$$\sigma_N^\infty = Y_1(2c/W)\sigma_N \tag{4.84}$$

The function Y_1 can be approximated by the expression[37]

$$Y_1(2c/W) = 1 + 0.128(2c/W) - 0.288(2c/W)^2 + 1.52(2c/W)^3 \tag{4.85}$$

This relationship is accurate within 0.5 percent for $2c/W \leq 0.7$. It has been shown by Cruse and Snyder[38] that the isotropic finite-width correction factor yields satisfactory results for most orthotropic plates of practical interest.

An unnotched tensile test yields σ_o which can be used in conjunction with σ_N^∞ and eq (2.144) to determine c_o. Since K_Q is not constant for many crack sizes of interest, it is necessary to use eq (2.144) to determine σ_N^∞ for other crack sizes. An average value of c_o, denoted by $\bar{c}_o$, can be obtained from eqs (2.144) and (2.145) in conjunction with experimentally measured notched strengths, σ_{Ni}, associated with n number of crack sizes, c_i, yielding the result

$$\bar{c}_o = \frac{1}{n}\sum_{i=1}^{n}\frac{c_i}{\left(\frac{\sigma_o}{\sigma_{Ni}}\right)^2 - 1} \tag{4.86}$$

This procedure yields a value of c_o such that an optimum fit is obtained between theory and experiment. Results of this procedure are shown in Fig. 4-39 for quasi-isotropic fiberglass (Scotchply 1002). Due to data scatter, variation in c_o with crack size is anticipated. For crack sizes of the same order as laminate thickness, free-edge effects and local heterogeneities often dominate the fracture process. In such cases c_o may be substantially different than for cracks which are large compared to the laminate thickness.

The stress fracture criteria require a value of σ_N^∞ for one crack length and a measurement of σ_o. Equations (2.154) and (2.159) can then be used to determine d_o and a_o, respectively. If the notched strength is obtained for a number of crack sizes, optimum values of these parameters, denoted by $\bar{d}_o$ and $\bar{a}_o$, can be determined from eq (4.86) in conjunction with eqs (2.165) and (2.166), yielding

$$\bar{d}_o = c_o/2, \qquad \bar{a}_o = 2c_o \tag{4.87}$$

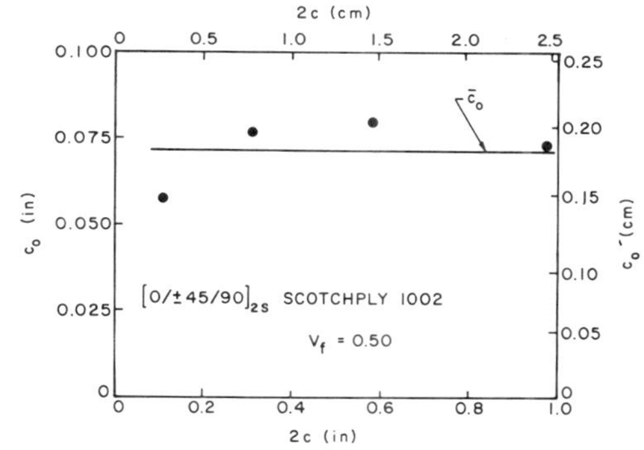

Fig. 4-39—Variation of inherent flaw size, c_o, with crack length (data from Nuismer and Whitney[35])

Comparison between theory and experiment for Scotchply 1002 quasi-isotropic laminates is shown in Figs. 4-40, 41, 42. Each data point represents an average of three specimens. A failed center-crack specimen is shown in Fig. 4-43. Note in Fig. 4-41 that K_Q does increase with crack size as predicted by the fracture criterion. As previously discussed, the point stress criterion does not yield a constant value of K_C for all crack sizes. The results in Fig. 4-42 illustrate, however, that the departure from a constant value of K_C is not of practical significance.

It has been shown by Morris and Hahn[39] that tensile coupons containing center cracks at an angle θ to the direction of the tensile load,

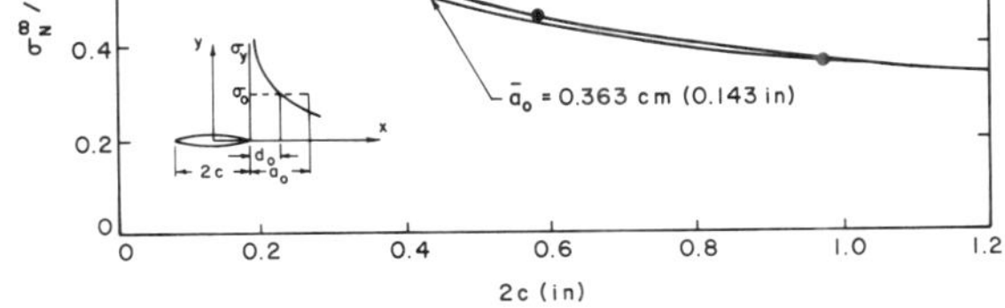

Fig. 4-40—Comparison between experimental data and stress fracture criterion for center crack (data from Nuismer and Whitney[35])

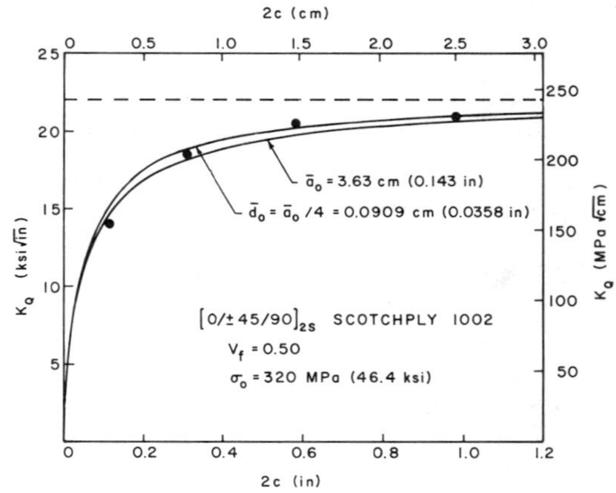

Fig. 4-41—Comparison between experimental data and predicted values of critical stress intensity factor, using the stress fracture criterion (data from Nuismer and Whitney[35])

see Fig. 4-44, can be handled in the same manner as a crack normal to the load by considering the projected length of the angle crack onto the direction normal to the load. In particular, the previous procedures for normal cracks can be utilized by considering an effective crack length, $\bar{c}$, defined by the relationship

$$\bar{c} = c \cos \theta \tag{4.88}$$

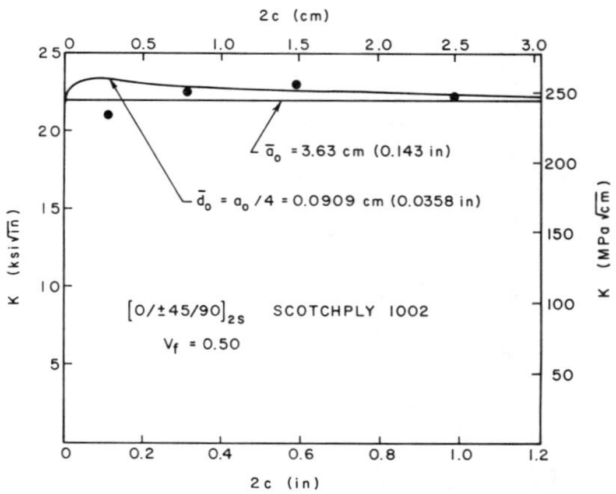

Fig. 4-42—Comparison between experimental data and predicted values of fracture toughness, using the stress fracture criterion (data from Nuismer and Whitney[35])

As in the case of the center crack, a straight-sided tensile coupon, as shown in Fig. 4-45, is utilized for a circular hole. The holes are formed by drilling mechanically or ultrasonically. A backing material should be used in conjunction with mechanical drilling procedures in order to prevent delamination of the bottom plies as the drill bit exits.

As in the case of the center crack, the notched strength, σ_N (failure load divided by unnotched cross-sectional area), is obtained for specimens containing circular holes. For applying the fracture mechanics approach to the problem of circular hole size effect in an isotropic plate, the following relationship can be used[40]

$$K_C = \sigma_N \sqrt{\pi a}\; F(2R/W, 2b/W) \tag{4.89}$$

where

$$b = R + a \tag{4.90}$$

The function $F(2R/W, 2b/W)$ is displayed graphically in Ref. 40. Equation (2.145), in conjunction with eq (4.89), yields the following results for an isotropic material

$$\sigma_N^\infty = Y_2(a/R, 2R/W, 2b/W)\sigma_N \tag{4.91}$$

where

$$Y_2(a/R, 2R/W, 2b/W) = \frac{F(2R/W, 2b/W)}{f(a/R)} \tag{4.92}$$

For orthotropic materials it is possible that Y_2 will also be a function of elastic properties.

For isotropic materials

$$K_T = Y_3(2R/W)K_T^\infty \tag{4.93}$$

For $2R/W \leq 1/3$, Y_3 can be very accurately approximated by the expression[41]

$$Y_3(2R/W) = \frac{2 + (1 - 2R/W)^3}{3(1 - 2R/W)} \tag{4.94}$$

If it is assumed that the stress concentration factor is of primary importance in determining the strength reduction in the case of a circular hole, then

Fig. 4-43—Failed center crack specimen, Scotchply 1002, $(0/\pm 45/90)_{2s}$ laminate (Nuismer and Whitney[35])

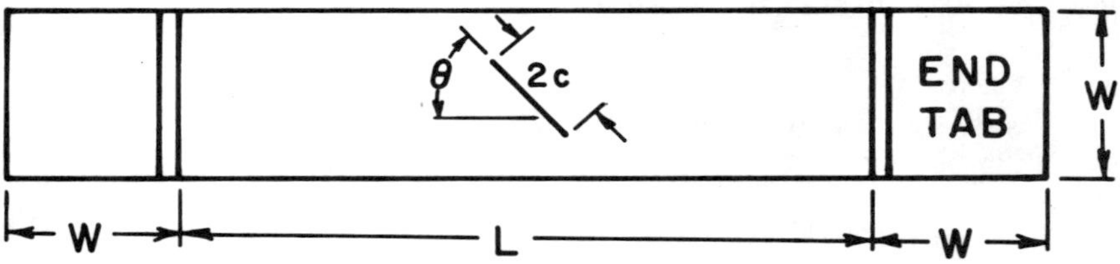

Fig. 4-44—Angle-crack specimen (Morris and Hahn[39])

$$\sigma_N^\infty = Y_3(2R/W)\sigma_N \tag{4.95}$$

Implicit in eq (4.95) in conjunction with the stress failure criterion is the assumption that the normal stress distribution adjacent to a circular hole in a finite-width orthotropic plate is equal to the stress distribution in an infinite plate multiplied by the isotropic finite-width correction factor. In Ref. 35 of Chapter 2, a constant-stress finite element program was used to generate the normal stress ahead of a hole in a finite-width plate of $2R/W = 1/3$ for a number of laminated composites. These results were plotted against the normal stress distribution obtained by multiplying the infinite plate stress by the isotropic finite-width correction factor, Y_3. A typical result is shown in Fig. 4-46 for the case of

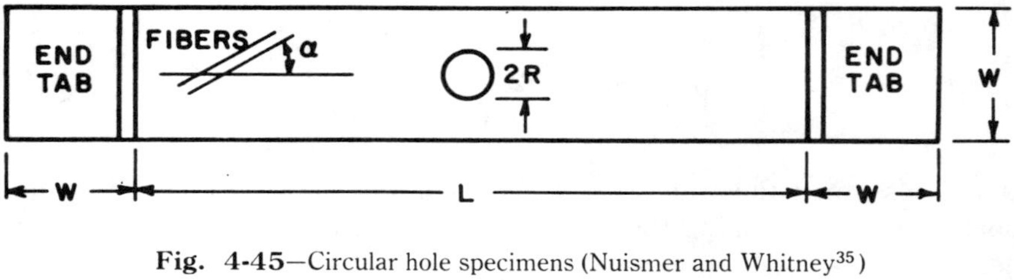

Fig. 4-45—Circular hole specimens (Nuismer and Whitney[35])

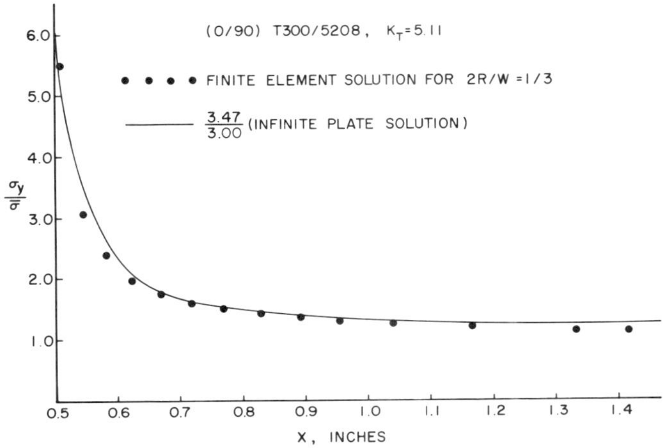

Fig. 4-46—Comparison of normal stress distributions across the ligament of an orthotropic finite-width plate containing a circular hole (Nuismer and Whitney[35])

a hole in a [0/90] graphite/epoxy plate where it is seen that the comparison is quite good.

Both the fracture mechanics approach and the stress fracture criterion require a measurement of the unnotched strength, σ_o, and the notched strength, σ_N^∞, for one hole size in order to determine the damage zone parameters a, a_o, and d_o. As in the case of the center crack, these parameters can be determined for a number of hole sizes such that theory and experiment coincide and then an average value used to develop an interpolation curve for any hole size. Since eqs (2.146), (2.150) and (2.158) cannot be solved explicitly for the damage zone parameters, an iteration process is necessary in conjunction with each measured hole size in order to match theory and experiment.

The variation of a_o with hole size is shown in Fig. 4-47 for a graphite/epoxy, [0/90] laminate. Note the large value of a_o associated with the one-inch diameter hole. It has been noted by Daniel[42] that large holes in certain laminate geometries produce local buckling which allows further distribution of damage, yielding a different failure process, and a higher notch strength than in the case of smaller hole sizes. Because of this effect, the average value of a_o, denoted by $\bar{a}_o$, is based on the three small hole sizes rather than all four hole sizes. As in the case of the center crack, it is anticipated that hole sizes of the same order as laminate thickness could produce a fracture process which is dominated by free-edge effects and local heterogeneities, resulting in a substantially different value of the damage zone parameters.

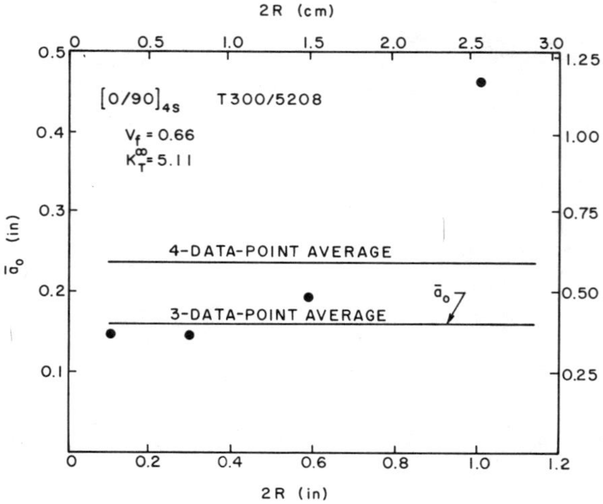

Fig. 4-47—Variation of damage zone parameters with hole size (data from Nuismer and Whitney[35])

A comparison between theory and experiment for graphite/epoxy laminates is shown in Fig. 4-48 for the stress fracture criterion and Fig. 4-49 for the fracture mechanics criterion. The bars over the damage zone parameters denote average values. In Fig. 4-48, $\bar{d}_o$ is also based on the three smaller holes. The data reduction procedure used in con-

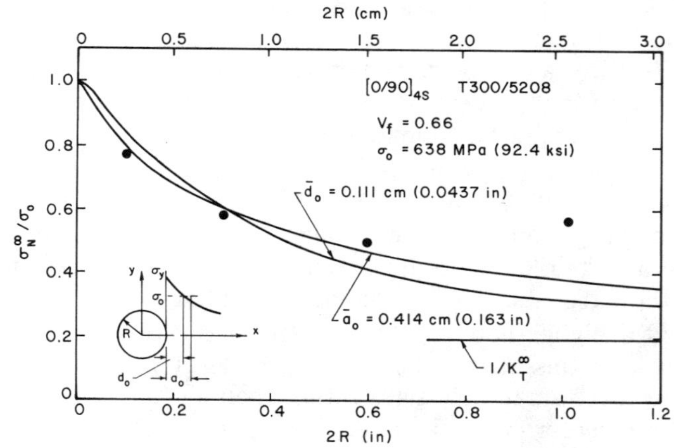

Fig. 4-48—Comparison between experimental data and stress fracture criterion for circular hole (data from Nuismer and Whitney[35])

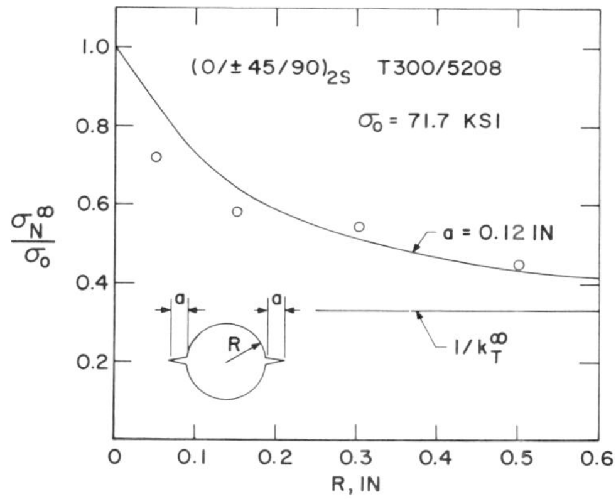

Fig. 4-49—Comparison between experimental data and fracture mechanics criterion for circular hole (data from Waddoups, Eisenmann and Kaminski[43])

junction with Fig. 4-49 is based on eq (4.88) rather than the more complex expression given by eq (4.91). In this procedure, as in the case of the stress fracture criterion, it is assumed that failure is dominated by the stress concentration factor. This procedure was also used in Ref. 43. All data points in Figs. 4-48 and 4-49 are based on an average of three specimens. Typical failed specimens are shown in Fig. 4-50.

4.4.2 First Ply Failure

In the determination of damage for laminated composites, particular attention is given to the effect of individual ply failure on the mechanical

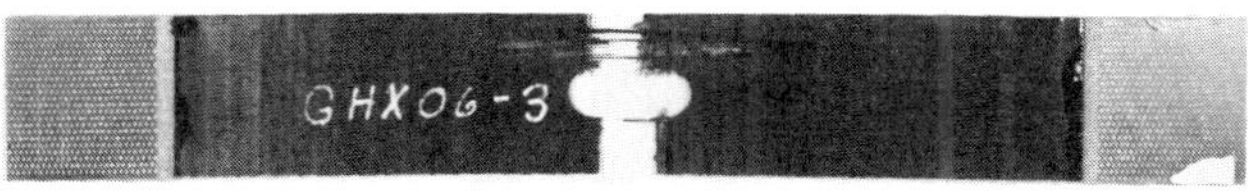

(a)

(b)

Fig. 4-50—Failed specimens, T300/5209 graphite/epoxy laminates with circular holes: (a) $[0/90]_{4s}$, (b) $[0/\pm 45/90]_{2s}$ (Nuismer and Whitney[35]) response of laminates, i.e., from an engineering point of view, does first ply failure constitute laminate failure? Individual ply failure is defined in terms of inplane, through cracks which would cause failure if the material were not an integral part of the laminate. In addition, the question of how to experimentally determine first ply failure is of concern.

Since 90-deg tension usually produces the lowest strain-to-failure in a unidirectional composite, first ply failure is of most concern in laminates which contain plies with the fibers oriented transverse to the primary load direction. For laminates in which the 90-deg plies carry substantial load, first ply failure can be experimentally observed by the presence of a knee, or plateau, in the longitudinal stress–strain curve. The stress–strain curves in Fig. 4-51 for glass-epoxy bidirectional laminates illustrates such behavior. A knee corresponding to 90-deg ply failure is observed in the case where half the plies are oriented at 0-deg. For the laminate in which only a quarter of the plies are at 0 deg, first ply failure produces a plateau in the stress–strain curve. The dominance of the 90-deg plies in this case also leads to significantly lower primary and secondary moduli compared to the laminate which has balance between the 0-deg and 90-deg plies.

In the case of the bidirectional graphite/epoxy laminate illustrated in Fig. 4-52, the longitudinal stress–strain curve for initial and subsequent loading does not show any effect of first ply failure. This is due to the fact that the 90-deg plies have very little influence on the longitudinal stress–strain curve due to the low value of E_1/E_2 for the unidirectional material. A transverse strain gage, however, produces a

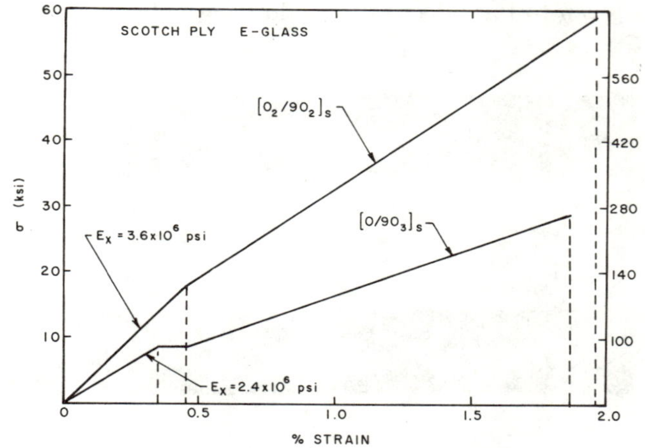

Fig. 4-51—Longitudinal stress–strain behavior of bidirectional glass/epoxy laminates under tension (Whitney, Browning and Grimes[44])

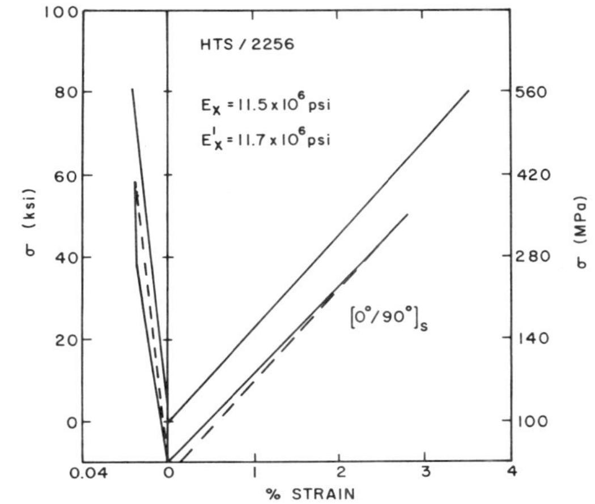

Fig. 4-52—Initial/subsequent tensile loading of bidirectional graphite/epoxy laminate (Whitney, Browning and Grimes[44])

knee in the initial stress–strain curve similar to the one observed in the longitudinal stress–strain response of the glass/epoxy laminate of Fig. 4-51. Microscopic examination of graphite/epoxy bidirectional laminates which have been loaded beyond the strain level which produces the knee in the transverse stress–strain curve reveals significant cracking.[13,44] The strain level at which both the knee in the stress–strain curves and the microcracking occurs, is approximately the strain level at which 90-deg failure occurs in a unidirectional composite. The onset of microcracking, however, occurs at strain levels below actual first ply failure.[45]

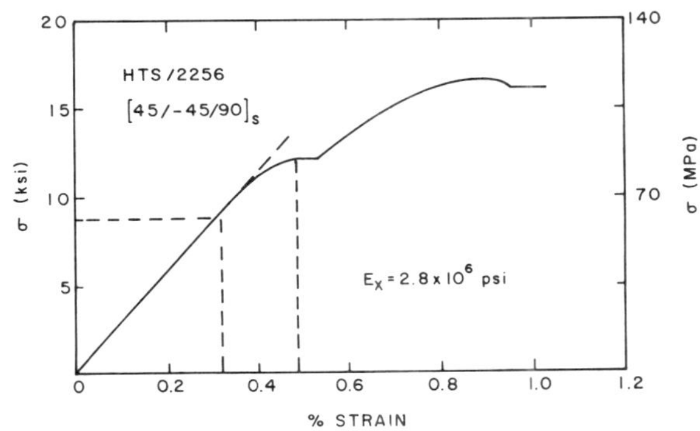

Fig. 4-53—Tensile stress–strain response of matrix dominated graphite/epoxy angle-ply laminate (Whitney, Browning and Grimes[44])

This initial cracking can be predicted from laminated plate theory in conjunction with the maximum strain criterion, provided residual stresses and strains due to laminate curing are accounted for.[45,46] Note in Fig. 4-52 that subsequent loading to failure after the initial loading does not produce a knee in the transverse stress–strain curve. A significant reduction in Poisson's ratio on subsequent loading is the only noticeable change in static properties.

In the case of $[\pm 45/90]_s$ laminates, shown in Figs. 4-53, 54, 55, failure of the 90-deg ply is directly reflected in the longitudinal stress–strain curve. In Fig. 4-53, 90-deg ply damage for graphite/epoxy is reflected by a plateau in the stress–strain curve. This type of behavior suggests that the failed ply unloads through a steep negative tangent modulus as discussed by Petit and Waddoups.[47] Initial and subsequent loading to a strain level below the plateau does not produce any measurable effect on the stress–strain curve, as shown in Fig. 4-54. For initial and subsequent loading above the plateau, however, Fig. 4-55, the first cycle produces significant hysteresis effects and a corresponding loss of tensile modulus. Further cycles to the same strain level produce elastic stress–strain response. No loss in ultimate strength is apparent due to the preloading cycles. It is anticipated, however, that successive loading to higher strain levels above the plateau would produce further hysteresis effects and degradation of tensile modulus.

In Fig. 4-56, the response of a graphite/epoxy $[0/\pm 45]_s$ laminate is shown. In the absence of 90-deg plies, ultimate failure is initiated by

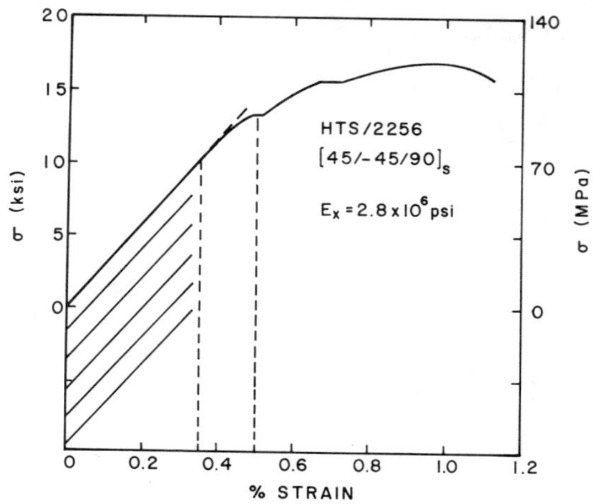

Fig. 4-54—Initial/subsequent tensile loading below first ply failure for matrix dominated graphite/epoxy angle-ply laminate (Whitney, Browning and Grimes[44])

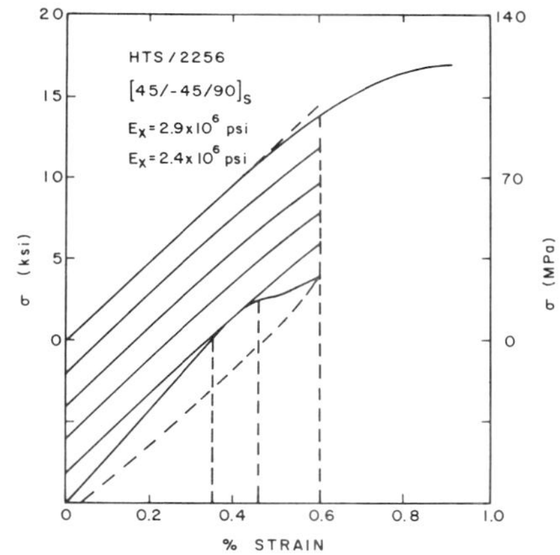

Fig. 4-55—Initial/subsequent tensile loading above first ply failure for matrix dominated graphite/epoxy angle-ply laminate

simultaneous destruction of the 0-deg and 45-deg plies. Since first ply failure produces laminate failure, it is anticipated that initial and subsequent tensile loading to a high strain level would produce little effect on the tensile stress–strain curve. Such a response is verified in Fig. 4-56.

Although the initial/subsequent tensile loads show composite damage with first ply failure, the significance of this damage is still not clear. For example, the plateau in the $[\pm 45/90]_s$ graphite/epoxy laminate of Fig. 4-53 can be removed by preloading the composite above first ply failure. As long as this strain level is not exceeded, the material appears

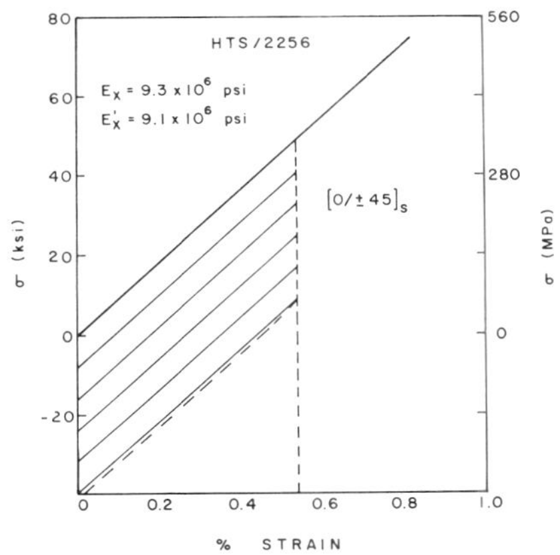

Fig. 4-56—Initial/subsequent tensile loading for filament dominated graphite/epoxy angle-ply laminate (Whitney, Browning and Grimes[44])

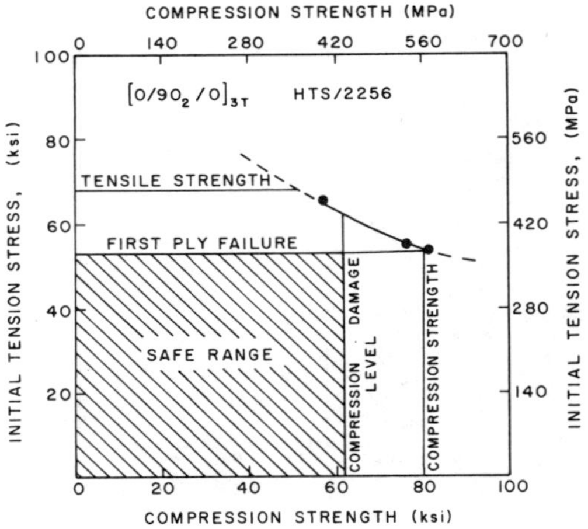

Fig. 4-57—Initial tension/subsequent compression strength of bidirectional graphite/epoxy laminate (Grimes et al[13])

to behave elastically. In order to determine significant damage, other loading conditions must be considered. In Fig. 4-57, the subsequent ultimate compressive strength is plotted as a function of an initial-tensile load for $[0/90_2/0]_{3T}$ HTS/2256 laminates. Tensile loads below first ply failure have little effect on the ultimate compressive strength. For increasing initial-tensile loads above first ply failure, there is, however, a corresponding loss in ultimate compressive strength. Thus, measurable damage is produced by first ply failure. A standard tensile coupon was used to perform the test, and platens were used for side support in order to guarantee stability under compressive loading. Further details can be found in Ref. 51.

Although some evidence has been presented to show that first ply failure can produce significant damage for certain laminates and certain loading sequences, the question as to whether first ply failure is a reasonable choice for limit and/or ultimate design strength is still open. The effect of first ply failure on such design considerations as fatigue life is not clear.

4.4.3 Biaxial Loading

As discussed in section 4.3.4, the off-axis tensile test can be used as a means of measuring biaxial stress–strain behavior relative to the fiber direction. This method could be extended to include orthotropic laminates in an off-axis orientation. This method, however, has a number of drawbacks with the most significant being the inability to independently control the biaxial stress components. Once a tensile test method is discarded, alternate approaches require special equipment and fixtures.

An ideal biaxial test method must meet the following requirements

(a) A significant volume of material must be under a homogeneous state of stress (uniform $\bar{\sigma}_x$, $\bar{\sigma}_y$, $\bar{\tau}_{xy}$).

(b) Primary failure must occur in the test section.

(c) The state of stress must be known without secondary measurements or analysis.

(d) It must be possible to vary the stress components independently.

A thin walled tubular specimen under combined axial load, torsion, and internal pressure meets the above requirements. As mentioned in section 4.3.6, this method is hindered by difficulties associated with fabricating quality tubular specimens and by difficulties associated with load introduction.

A common method for fabricating tubular specimens utilizes a steel cavity tool, hollow, perforated mandrel, and elastomer bladder. The tubular laminate is internally pressurized so that it conforms to the shape of the cylindrical cavity. Complete details on such a procedure can be found in Refs. 48 and 49. Fiber volume content and uniformity of wall thickness are two of the most difficult parameters to control.

Axial loads and internal pressure present a particular problem in terms of end attachments. In particular, when the transverse and radial displacements of a tubular specimen are restricted by end attachments, extraneous bending stresses are induced in tubular regions near the end attachments. Such stress concentrations have been confirmed analytically and experimentally[50] for tubular specimens under axial load. Laminates which have very large effective Poisson's ratios, such as $[\pm 45]_s$ laminates, are a particular problem. These stresses and strain concentrations may induce premature end failures, and thus render the ultimate-strength results meaningless. Bonded grips, as discussed in section 4.3.4, allow some radial dimensional change, and as a result can be utilized for many laminates and load combinations. Self-compensating end grips have also been developed.[49,51] Such designs utilize combinations of internal and external pressure in the end grip area to provide radial displacements which match gage-section displacements. Strain gages can be used to sense any strain gradient and automatically trigger the self-compensating device. Grips of this nature are very expensive and require peripheral electronic equipment.

Consider a tube with a cartesian coordinate system located in the middle surface of the tube with x, y, and z measured along the longitudinal circumferential, and radial directions, respectively. For any combination of axial load, internal pressure, and torque, the average stresses in the tube are given by

$$\bar{\sigma}_x = \frac{N_x}{h}, \qquad \bar{\sigma}_y = \frac{N_y}{h} = \frac{R}{h}p,$$

$$\bar{\tau}_{xy} = \frac{T}{2\pi h R^2} \tag{4.96}$$

where h is the thickness of tube wall, R is the radius of the middle surface of the tube, p is the internal pressure, and T is the torque. The constitutive equations of the cylinder are obtained from eq (2.83) for the case where the moments vanish, with the result

$$\epsilon_x^\circ = H_{11}h\bar{\sigma}_x + H_{12}Rp + H_{16}\frac{T}{2\pi R^2} \tag{4.97}$$

$$\epsilon_y^\circ = H_{12}h\bar{\sigma}_x + H_{22}Rp + H_{26}\frac{T}{2\pi R^2} \tag{4.98}$$

$$\gamma_{xy}^\circ = H_{16}h\bar{\sigma}_x + H_{26}Rp + H_{66}\frac{T}{2\pi R^2} \tag{4.99}$$

Effective elastic constants can be determined from H_{ij} by applying eqs (2.84–2.88).

For anisotropic tubes (H_{16}, $H_{26} \neq 0$), eq (4.99) reveals that a shear strain will be induced by axial tension. In such cases one end of the tube must be allowed to rotate or shear stress is induced. Under certain circumstances shear buckling can result.[52] A rectangular, three-element rosette gage can be used to measure outer surface strains which are related to midplane strains in the following manner.[53]

$$\epsilon_x^\circ = \epsilon_x \tag{4.100}$$

$$\epsilon^\circ = \frac{\epsilon_y}{1 + h/2R} \tag{4.101}$$

$$\gamma_{xy}^\circ = \frac{\gamma_{xy}}{1 + h/2R} \tag{4.102}$$

Dimensions of the tube should be chosen to minimize strain gradient effects and to assure uniform stresses in the gage section. The former can be accomplished by making $R/h \geq 10$, and the latter can be assured by utilizing the realtionship[28 54]

$$L = 4R + \bar{L} \tag{4.103}$$

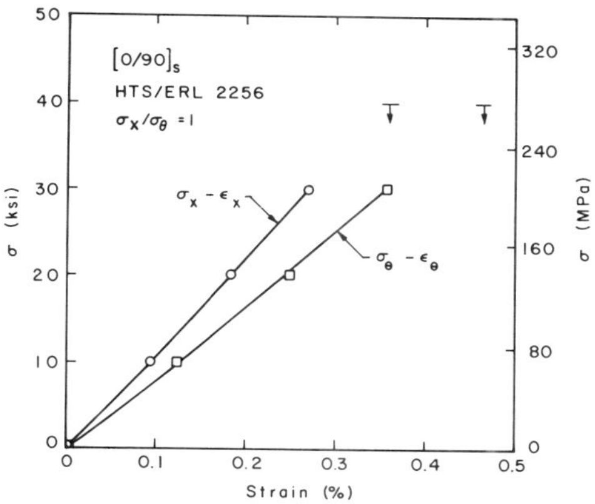

Fig. 4-58—Stress–strain curves for combined longitudinal tension (LT) and internal pressure (IP) loading of a bidirectional graphite/epoxy laminated tube (Grimes and Francis[30])

where L denotes total tube length and $\overline{L}$ denotes the desired tube gage length.

Stress–strain curves are shown in Figs. 4-58 and 4-59 for combined loading of a $[0/90]_s$ laminated tube fabricated from Hercules' HTS

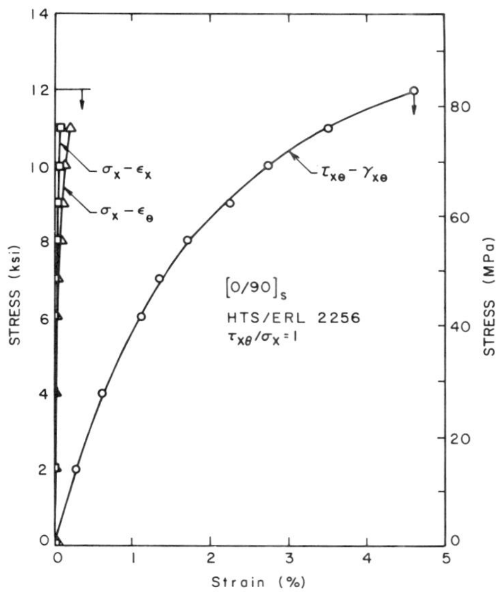

Fig. 4-59—Stress–strain curves for combined longitudinal tension (LT) and torsion (T) loading of a bidirectional graphite/epoxy laminate tube (Grimes and Francis[30])

graphite fiber in Union Carbide's ERL 2256 epoxy resin.[30] Longitudinal tension (LT) and internal pressure (IP), one-to-one ratio, are shown in Fig. 4-58. Combined longitudinal tension and torsion (T), one-to-one ratio, are shown in Fig. 4-59. Note that in Fig. 4-59, the longitudinal tension has little effect on the shear stress–strain response. This can easily be seen by comparing Fig. 4-59 with the pure shear results shown in Fig. 4-35.

While the tubular specimen is ideal for measuring combined load response for unnotched composites, they are unacceptable for determining the biaxial response of notched composites. Evaluation of composites containing stress concentrations, such as circular holes, slits, and cutouts, requires use of flat plate specimens. For combined loading, such an evaluation requires rather large specimens and specialized test fixtures. A flat plate biaxial test has been developed by Daniel.[22,42]

This method utilizes 8-ply specimens 40 cm × 40 cm (16 in. × 16 in.). Corners of approximately 5 cm (2 in.) sides are cut off from the composite laminates. They are then tabbed with 5-ply cross-ply glass-

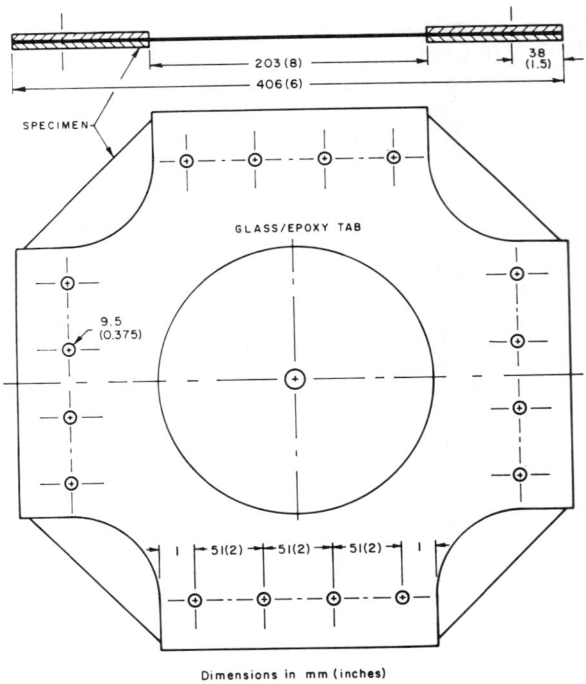

Fig. 4-60—Sketch of biaxial test specimen, dimensions in mm and inches (Daniel[42])

Fig. 4-61—Whiffle-tree linkage grips for load introduction in biaxial specimen (Daniel[42])

epoxy tabs. These tabs contain a circular cutout in the center of 20.3 cm (8 in.) diameter. A sketch of this specimen is shown in Fig. 4-60.

Deformations and strains are measured using strain gages. For Daniel's studies,[22,42] gages were mounted on the hole boundary, near it, and in the far field along the horizontal (*x*-axis) and vertical (*y*-axis) of material symmetry. In most cases birefringent coatings 0.5 mm (0.02 in.) and 1 mm (0.04 in.) thick were used on one side of the specimen.

Four 0.95 mm (0.037 in.) diameter holes are provided on each side of the tabbed specimen for bolting individual pairs of metal grips approximately 5 cm (2 in.) wide and 10 cm (4 in.) long. Loading is introduced by means of four whiffle-tree grip linkages designed to ensure that four equal loads are applied to each side of the specimen. A photograph of a biaxial specimen with the loading grip linkages is shown in Fig. 4-61. Note that the tab geometry and slits are from an early design which was found to be unacceptable due to stress concentrations. The grip linkage concept, however, was retained in later designs.

Load is applied by means of two pairs of hydraulic jacks attached to the four sides of a reaction frame. The load is transmitted from the hydraulic cylinders to the grip linkages through the bore of these cylinders. The rods are instrumented with strain gages and calibrated in a testing machine to establish the exact relationship between loads and strain gage signals. These strain gage readings are used subsequently both for recording the exact loads applied to the specimen and as feedback signals for controlling the pressures by means of the servohydraulic systems used.

The relationship between the input load and the effective stress for the specimen geometry used and stress biaxiality desired must be established by preliminary calibration tests. Unnotched specimens tabbed as previously described and instrumented with strain gages at the center and far-field locations can be used to perform the calibration. Using eq (2.84) for the case of an orthotropic laminate ($H_{16} = H_{26} = 0$) in conjunction with the definition of elastic constants, eqs (2.79–83), one obtains

$$\epsilon_x = \frac{N_y}{hE_x}(B - \nu_{xy}) \tag{4.104}$$

$$\epsilon_y = \frac{N_y}{hE_x}\left(\frac{E_x}{E_y} - \nu_{xy}B\right) \tag{4.105}$$

where B is the biaxiality ratio N_x/N_y. Combining eqs (4.104) and (4.105) yields

$$\frac{\epsilon_x}{\epsilon_y} = \frac{(B - \nu_{xy})}{(E_x/E_y - \nu_{xy}B)} \tag{4.106}$$

Substituting measured values of the material constants E_x, E_y, and ν_{xy} along with the desired value of B into eq (4.106) yields ϵ_x/ϵ_y. The input load ratio must be adjusted until the desired strain ratio in eq (4.106) is obtained at the center of the plate.

Stress–strain response for a $[90_2/\pm 45]_s$ graphite/epoxy laminate containing a 2.54 cm (1 in.) circular hole and subjected to a biaxial load, $B = 0.5$, is shown in Figs. 4-62 and 4-63.[42] The laminate is fabricated

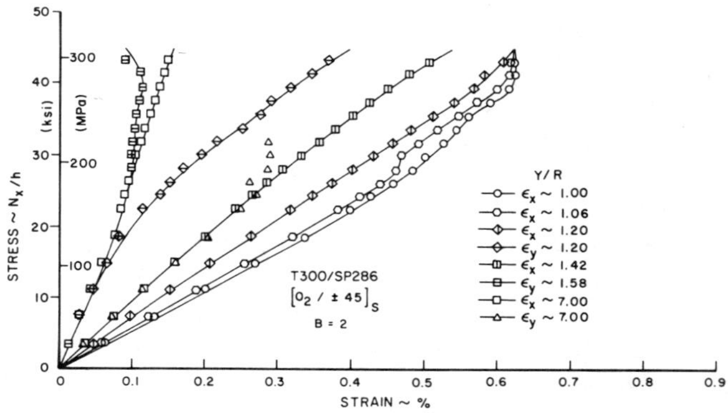

Fig. 4-62—Strains along horizontal axis of graphite/epoxy laminate with 2.54 cm (1 in.) diameter hole under biaxial loading (Daniel[42])

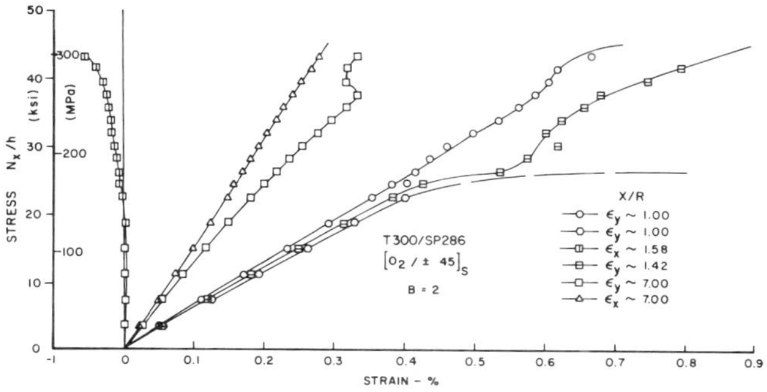

Fig. 4-63—Strains along vertical axis of graphite/epoxy laminate with 2.54 cm (1 in.) diameter hole under biaxial loading (Daniel[42])

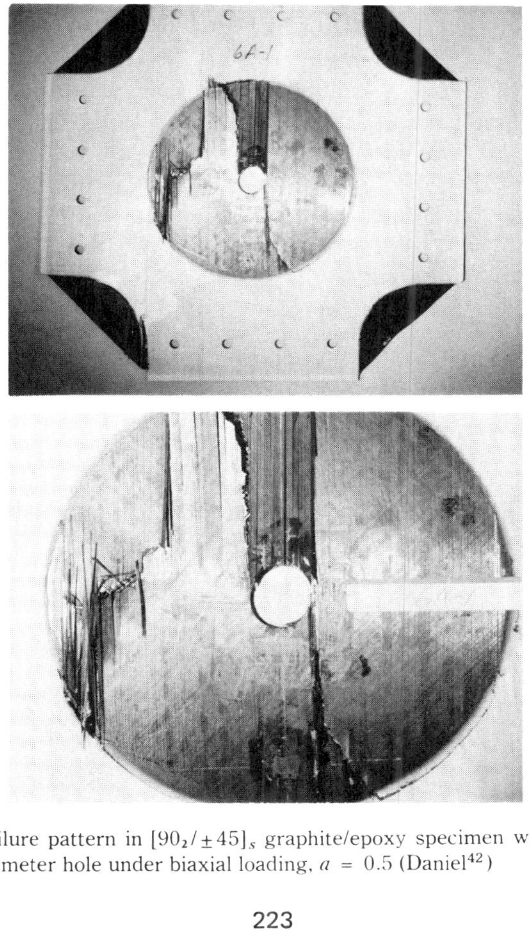

Fig. 4-64—Failure pattern in $[90_2/\pm 45]_s$ graphite/epoxy specimen with 2.54 cm (1 in.) diameter hole under biaxial loading, $a = 0.5$ (Daniel[42])

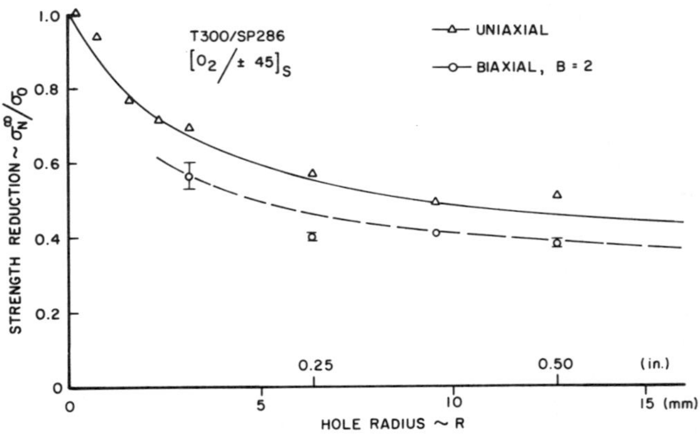

Fig. 4-65—Strength reduction as a function of hole radius for graphite/epoxy plates with circular holes under uniaxial and biaxial loading (Daniel[42])

from T300/SP286 pre-preg material manufactured by 3M Company. The failure pattern for this laminate is shown in Fig. 4-64. A comparison between the biaxial strength and uniaxial strength of this laminate is shown in Fig. 4-65 for various hole sizes. Note that the biaxial data follow a hole-size trend similar to the uniaxial data, but the biaxial data have larger strength reductions.

For laminates containing a stress concentration in the center, the flat plate biaxial test method described appears to yield excellent results. Difficulties are encountered with unnotched specimens, however, as stress concentrations associated with load introduction may lead to pre-mature failure.

4.4.4 Fatigue Loading

Fatigue testing of composite materials is a complex subject. Choice of load history, specimen design, and interpretation of data present difficulties. Considerable research effort is being concentrated in the area of fatigue testing and design. Thus the state-of-the-art in fatigue testing of composite materials is in its infancy.

The case of constant-amplitude tension–tension fatigue of fiber reinforced composites has received considerable attention which has led to the development of a test method utilizing the straight-sided tensile coupon described in section 4.3.1. Complete details of the method are described in ASTM D3479-76. As in the case of the tensile test, the method is applicable to any orthotropic ($A_{16} = A_{26} = 0$), symmetric ($B_{ij} = 0$) laminate. The test method has two options involving either constant stress amplitude or constant strain amplitude.

When utilizing this test method one must be aware of a number of basic characteristics of polymeric matrix composites which may influence fatigue life. Many resin systems display appreciable damping and as a result any temperature rise during fatigue testing can affect life. High frequency testing can induce considerable temperature rise. Thus, frequency dependence is often observed. Fatigue damage, such as resin micro-cracking, can develop without visual evidence, leading to loss in stiffness and residual strength. Under certain circumstances such changes may constitute failure rather than total fracture. In addition, a complex three-dimensional state of stress exists near the free edge of laminates, as discussed in section 2.9, which can induce delamination under fatigue loading.[15] Because of free-edge delamination it is difficult to relate laboratory coupon fatigue test results to the behavior of structures which may or may not have free edges. Thus, size effect can be an even more important consideration in the fatigue design for composites than it is for metals.

As in the case of the standard tensile test, strain gages or an extensometer can be used to monitor strain. If any chance exists for specimen heating, temperature should be monitored during the fatigue test.

Since fatigue damage can take different forms it is important that a definition of failure is clearly stated when reporting fatigue life data. Any observed preliminary damage should also be reported. Such information can be useful in assessing fatigue damage mechanisms and in developing fatigue failure models.

Presentation of data is also an important consideration in fatigue testing. The classical S–N curve is a primary method of characterizing fatigue behavior. This method consists of determining the number of cycles to failure at a number of stress levels. The resulting S–N curve yields an estimate of the mean time-to-failure as a function of stress level. Such a procedure, however, fails to account for the large variation in the time-to-failure at a given stress level. In particular, when the mean time-to-failure has been reached, half of the samples have failed. From a design standpoint, one is interested in survivability above 90 percent.

Fatigue data with statistical significance requires a large number of replicates at a given stress level in order to measure the distribution of time-to-failure. The "wearout" model approach[5,7,55] provides one means of generating an S–N curve with some statistical value without resorting to an extremely large data base. Such an approach, however, involves the assumption of a direct relationship between static strength distribution, residual strength distribution after time under a specified load history, and distribution of time-to-failure at a maximum stress level. Since the mechanisms of failure under fatigue loading are not well understood, these assumed relationships may be premature.

An alternative to the "wearout" model approach has been proposed by Hahn and Kim.[56] The approach is based on two assumptions: (1) a classical power-law representation of the S–N curve and (2) a two-parameter Weibull distribution of time-to-failure. In mathematical form these assumptions become

$$NS^{b} = C \tag{4.107}$$

where S is the maximum stress, N is the number of cycles to failure, and b and c are material constants. In addition,

$$R(N) = \exp\left[-\left(\frac{N}{\overline{N}}\right)^{\alpha_f}\right) \tag{4.108}$$

Here $R(N)$ is the probability of survival for N cycles, where $\overline{N}$ is the characteristic time to failure, and α_f is the fatigue shape parameter. Equations (4.107) and (4.108) can also be obtained from the "wearout" model approach with α_f being related to the shape parameter for static strength.[55] Combining eqs (4.107) and (4.108) yields

$$S = K\{[-\ln R(N)]^{-1/\alpha_f b}\}\overline{N}^{-1/b} \tag{4.109}$$

where

$$K = C^{1/b} \tag{4.110}$$

When $N = \overline{N}$, $-\ln R(\overline{N}) = 1$ and eq (4.110) reduced to

$$S(\overline{N}) = K\,\overline{N}^{-1/b} \tag{4.111}$$

A plot of log S versus log $\overline{N}$ produces a straight line with slope $-1/b$ and a y intercept of log K. Thus, a measure of the distribution of time-to-failure at various stress levels in conjunction with eq (4.108) allows α_f to be determined along with a set of values of $\overline{N}$, each corresponding to a value of S.

An S–N curve for any desired level of reliability can be obtained by writing eq (4.111) in the form

$$\overline{N} = \left(\frac{K}{S}\right)^{b} \tag{4.112}$$

and substituting into eq (4.108) with the result

$$R(N) = \exp\left[-\left(\frac{N}{C^{-1}S^{-b}}\right)^{\alpha_f}\right] \tag{4.113}$$

Solving for S leads to the following S–N relationship for any desired level of reliability

$$S = K\{[-\ln R(N)]^{1/\alpha_f b}\}N^{-1/b} \tag{4.114}$$

The data reduction procedure can now be summarized as follows:

(a) Fit the time-to-failure data at each stress level to a two-parameter Weibull distribution.

(b) Use a data pooling scheme to determine the fatigue shape parameter α_f.

(c) Fit log S versus log $\overline{N}$ data to a straight line for the determination of b and K.

Let m be the number of stress levels tested and n_i the number of specimens tested at the i-th stress level, S_i, which leads to the data set

$$N_i(N_{i1}, N_{i2}, \ldots, N_{in_i}), \qquad i = 1, 2, \ldots, m \tag{4.115}$$

Each stress level is fit to the two-parameter Weibull distribution

$$R(N_i) = \exp\left[-\left(\frac{\mathrm{N}_i}{\overline{N}_i}\right)^{\alpha_{fi}}\right] \tag{4.116}$$

A number of procedures can be utilized for determining α_{fi} and $\overline{N}_i$. One of the methods preferred by statisticians is the maximum likelihood estimator (MLE) which is of the form[57]

$$\frac{\sum_{j=1}^{n_i} N_{ij}^{\alpha_{fi}} \ln N_{ij}}{\sum_{j=1}^{n_i} N_{ij}\,\hat{\alpha}_{fi}} - \frac{1}{n_i}\sum_{j=1}^{n_i} \ln N_{ij} - \frac{1}{\hat{\alpha}_{fi}} = 0 \tag{4.117}$$

$$\hat{N}_i = \left(\frac{1}{n_i}\sum_{j=1}^{n_i} N_{ij}^{\hat{\alpha}_{fi}}\right)^{1/\hat{\alpha}_{fi}} \tag{4.118}$$

where $\hat{\alpha}_{fi}$ and $\hat{N}_i$ denote estimated values of α_{fi} and $\overline{N}_i$, respectively. Equation (4.117) has only one real positive root. As a result, an iterative scheme can be utilized until a value of $\hat{\alpha}_{fi}$ is obtained to any desired number of decimal places. The resulting value of $\hat{\alpha}_{fi}$ obtained from the iterative scheme can then be used in conjunction with eq (4.118) to obtain $\hat{N}_i$.

Table 4-2 Equality of Weibull Shape Parameters, $\gamma = 0.98$ (Ref. 58)

n_i	A
5	3.550
10	2.213
15	1.870
20	1.703
100	1.266

Since estimated fatigue shape parameter, $\hat{\alpha}_{fi}$, are random variables based on a specific sample size, it is anticipated that each value of S_i would produce a different value of $\hat{\alpha}_{fi}$ even though α_f may be independent of stress level. A two-sample test[58] is available, however, which allows for testing the equality of shape parameters in two-parameter Weibull distributions with unknown scale parameters. The approach is based on MLE and the results depend on sample size and confidence level desired. Let $\hat{\alpha}_{max}$ and $\hat{\alpha}_{min}$ be the maximum and minimum values obtained for $\hat{\alpha}_{fi}$. For the information tabulated in Ref. 58, it is required that α_{max} and α_{min} be associated with equal sample sizes, n. If $\alpha_{f\,max}$ and $\alpha_{f\,min}$ are from the same distribution, then it is expected that[58]

$$\frac{\hat{\alpha}_{f\,max}}{\hat{\alpha}_{f\,min}} < A\,(\gamma, n), \qquad A > 1 \tag{4.119}$$

for a given confidence level, γ, and sample size, n. Values of A are shown in Table 4-2 for various sample sizes corresponding to a confidence level of 0.98. These data are taken from Ref. 58. The large values of A associated with small sample sizes suggest that significant variations in α_{fi} are likely to be encountered with small data sets taken from the same population.

Let us now assume that α_f is independent of stress level. Then a data pooling technique must be utilized in order to determine a single value of α_f for all S_i. Various approaches for obtaining a pooled value of α_f can be found in the literature. The approach used in the present work has been investigated by Wolff and Lemon.[59] This procedure utilizes the normalized data set

$$X(X_{i1}, X_{i2}, \ldots, X_{in_i}), \qquad i = 1, 2, \ldots, m \tag{4.120}$$

where

$$X_{ij} = \frac{N_{ij}}{\hat{N}_i} \tag{4.121}$$

Thus, each set of data at a given stress range is normalized by the estimated characteristic time-to-failure and the results fit to the pooled two-parameter Weibull distribution

$$R(X) = \exp - [(\frac{X}{X_o})^{\alpha_f}] . \tag{4.122}$$

This procedure has the advantage of obtaining a large data base for determining α_f by using a few replicates for a number of values of S_i. In general, for equal accuracy, fewer specimens are needed to determine the location parameter than to determine the shape parameter.

For the pooled Weibull distribution, eq (4.122), the MLE relationships take the form

$$\frac{\sum_{i=1}^{m} \sum_{j=1}^{n_i} X_{ij}^{\bar{\alpha}_f} \ell n X_{ij}}{\sum_{i=1}^{m} \sum_{j=1}^{n_i} X_{ij} \bar{\alpha}_f} - \frac{1}{M} \sum_{i=1}^{m} \sum_{j=1}^{n_i} \ell n X_{ij} - \frac{1}{\bar{\alpha}_f} = 0 \tag{4.123}$$

$$\overline{X}_o = (\frac{1}{M} \sum_{i=1}^{m} \sum_{j=1}^{n_i} X_{ij}^{\bar{\alpha}_f})^{1/\bar{\alpha}_f} \tag{4.124}$$

where $\bar{\alpha}_f$ and $\overline{X}_o$ are estimated values of α_f and X_o, respectively, and

$$M = \sum_{i=1}^{m} n_i \tag{4.125}$$

For a perfect fit to the data pooling scheme, the location parameter, X_o, should be unity. The value of $\hat{N}_i$ can be adjusted to produce an exact value of unity for X_o. In particular,

$$\overline{N}_{oi} = \overline{X}_o \hat{N}_i \tag{4.126}$$

where $\overline{N}_{oi}$ denotes estimated values of $\overline{N}$ associated with the adjusted two-parameter Weibull distribution

$$R(X) = \exp(-X^{\alpha_f}) \tag{4.127}$$

The slope of the S–N curve, $1/b$, and the y-intercept, K, can be determined by fitting log S_i versus log $\overline{N}_{oi}$ to a straight line. With K, b, and α_f now determined, eq (4.114) can be used to produce an S–N curve of any desired reliability.

It should be noted that MLE is asymptotically unbiased, i.e., it is a biased estimator for small sample sizes.[57] Unbiasing factors are tabulated in Ref. 60. These factors are less than unity as MLE always tends to overestimate the shape parameter. Confidence intervals for both the shape parameter and location parameter as a function of sample size have been established.[60] If conservative estimates are desired for $R(N)$, then a lower bound value of α_f can be utilized. This value of α_f can be used in conjunction with eq (4.124) to determine $\overline{X}_o$.

In the case of high-cycle fatigue the time-to-failure may become unacceptably long. This difficulty can be overcome by raising the stress level so that fatigue failures are produced within a reasonable number of cycles. For filament dominated laminates, which tend to have a very flat S–N curve, stress levels may have to be raised to an unacceptable level to produce a reasonable time-to-failure for all specimens tested. In particular, it is undesirable to raise the fatigue stress level to such a degree that it significantly overlaps the static strength distribution. In such cases the probability of a first cycle failure is significant.

If censoring techniques[57] are applied to data reduction procedures, fatigue failures are not required of all specimens. For the data reduction scheme outlined in the present work, Type I censoring seems to be the most desirable in terms of yielding the most information. In the case of Type I censoring, the fatigue test is terminated at a pre-determined time (e.g., 10^6 cycles) even though all specimens may have not failed. The MLE equations for Type I censoring are of the form[57,61]

$$\frac{\sum_{j=1}^{r_i} N_{ij}^{\hat{\alpha}_{fi}} \ell n\, N_{ij} + (n_i - r_i) R_i^{\hat{\alpha}_{fi}} \ell n\, R_i}{\sum_{j=1}^{r_i} N_{ij}\, \hat{\alpha}_{fi} + (n_i - r_i) R_i\, \hat{\alpha}_{fi}} - \frac{1}{r_i} \sum_{j=1}^{r_i} \ell n\, N_{ij} - \frac{1}{\hat{\alpha}_{fi}} = 0 \tag{4.128}$$

$$\hat{N}_i = \left\{ \frac{1}{r_i} \left[\sum_{j=1}^{n_i} N_{ij}^{\hat{\alpha}_{fi}} + (n_i - r_i) R_i^{\hat{\alpha}_{fi}} \right] \right\}^{1/\hat{\alpha}_{fi}} \tag{4.129}$$

where n_i now denotes the total number of specimens tested at S_i, r_i is the number of fatigue failures at S_i, and R_i is the number of cycles at which the test is terminated.

The data pooling procedure is now analogous to "progressive censoring" in which a number of samples are removed at pre-determined time intervals throughout the duration of the test. The MLE associated with the data pooling procedure in conjunction with censored samples becomes[61]

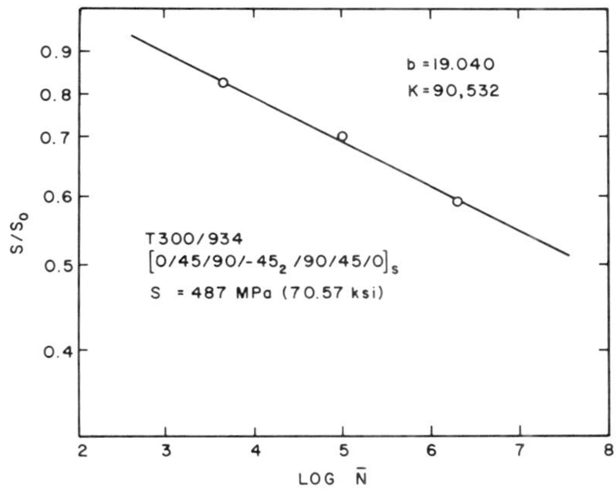

Fig. 4-66—Determination of S–N parameters for graphite/epoxy quasi-isotropic laminate (data from Ryder and Walker[15])

$$\frac{\sum_{i=1}^{m}\sum_{j=1}^{r_i} X_{ij}^{\bar{\alpha}_f} \ell n X_{ij} + \sum_{i=1}^{m} (n_i - r_i) Y_i^{\bar{\alpha}_f} \ell n y_i}{\sum_{i=1}^{m}\sum_{j=1}^{r_i} X_{ij}^{\bar{\alpha}_f} + \sum_{i=1}^{m} (n_i - r_i) Y_i^{\bar{\alpha}_f}} \tag{4.130}$$

$$- \frac{1}{F} \sum_{i=1}^{m}\sum_{j=1}^{r_i} \ln X_{ij} - \frac{1}{\bar{\alpha}_f} = 0 \tag{4.131}$$

$$X_o = \{\frac{1}{F} [\sum_{i=1}^{m}\sum_{j=1}^{r_i} X_{ij}^{\bar{\alpha}_f} + \sum_{i=1}^{m} (n_i - r_i) Y_i^{\bar{\alpha}_f}]\}^{1/\bar{\alpha}_f}$$

where

$$Y_i = \frac{R_i}{\bar{N}_i} \tag{4.132}$$

and F is the total number of fatigue failures, i.e.,

$$F = \sum_{i=1}^{m} r_i \tag{4.133}$$

To illustrate the fatigue characterization procedure, consider the tension–tension fatigue data on quasi-isotropic graphite-epoxy laminates generated by Ryder and Walker.[15] The laminates were fabricated from Narmco's T300/934 graphite-epoxy pre-preg system. Twenty replicates were fatigued to failure at each of three different stress levels. Three replicates at each of three additional stress levels were fatigued to

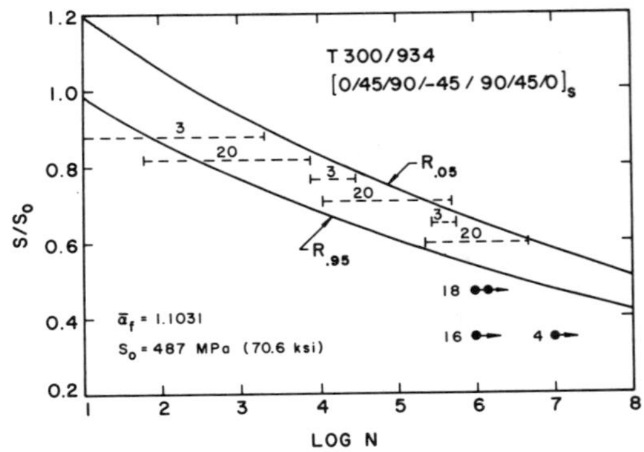

Fig. 4-67—S–N curve for graphite/epoxy quasi-isotropic laminate (data from Ryder and Walker[15])

failure as part of an initial stress-life scan. Results of the data reduction procedure are shown in Figs. 4-66 and 4-67 where S_o is the characteristic static strength. The data reduction procedure only included the three stress levels in which twenty replicates were tested. The S–$\overline{N}$ plot for determining b and K is shown in Fig. 4-66, while the resulting S–N curves associated with 5 percent and 95 percent probability of survival are shown in Fig. 4-67. Individual Weibull parameters for the three stress levels are shown in Table 4-3.

Scatter bands for all stress levels in which data were obtained are also shown in Fig. 4-67. For lower stress levels, which are of interest to designers, the data reduction procedure appears to yield good results. At higher stress levels the scatter bands are larger and fall outside of the 95 percent survivability curve. Data at higher stress levels tend to become less accurate as the tail of the static strength population is approached. A number of fatigue run-out data points are also shown in Fig. 4-67.

Table 4-3 Weibull Parameters

S MPa (ksi)	S/S_o	$\overline{N}$ Cycles	α_f
290 (42)	0.5951	2,006,599	1.3454
345 (50)	0.7085	101,024	1.2626
400 (58)	0.8219	4,322	0.7583
	Pooled	0.9926	1.1031

This procedure as outlined here provides a method for comparing data from different materials, load histories, and laminate orientations. An estimation of the endurance limit (value of S corresponding to $N = 10^7$ cycles) at any desired probability of survival can be obtained with this procedure. The value of b, which characterizes the slope of the S–N curve, and α_f, which characterizes the fatigue scatter are the key parameters of interest.

For more complex load histories, forms other than eq (4.107) can be used for the S–N relationship.

4.5 INTERLAMINAR FRACTURE MECHANICS CHARACTERIZATION

The only test methods discussed to this point which evaluate interlaminar strength of composite materials are the interlaminar beam tests presented in section 4.3.7. Delamination has received increasing recognition as a major failure mode in laminated fiber reinforced composite materials. Such failures are of particular importance in high-performance applications such as aircraft structures. Manufacturing defects and in-service induced damage, such as impact, are sources of delamination.

Although considerable research is currently being accomplished in the area of delamination, two test methods have emerged which appear to have high potential for delamination characterization. Both methods are based on concepts of classical linear elastic fracture mechanics (LEFM) with critical strain energy release rate being the characterization parameter.

The two delamination failure modes of major concern are the opening or peel mode (Mode I) associated with interlaminar normal stress, and the sliding shear mode (Mode II) associated with interlaminar shear stress. Mode I characterization can be accomplished utilizing the double cantilever beam (DCB) test and Mode II characterization by the end notch flexure (ENF) test. These are the two interlaminar fracture tests discussed. The concept of strain energy release rate will be briefly discussed prior to presentation of the DCB and ENF tests.

4.5.1 *Strain Energy Release Rate*

Consider a cracked plate of thickness h subjected to an external load P, as illustrated in Fig. 4-68. A crack extension from a to $a + da$ induces a change in compliance which results in a loss of strain energy, dU. Since this behavior involves the opening mode, we define the Mode I strain energy release rate, G_I, as

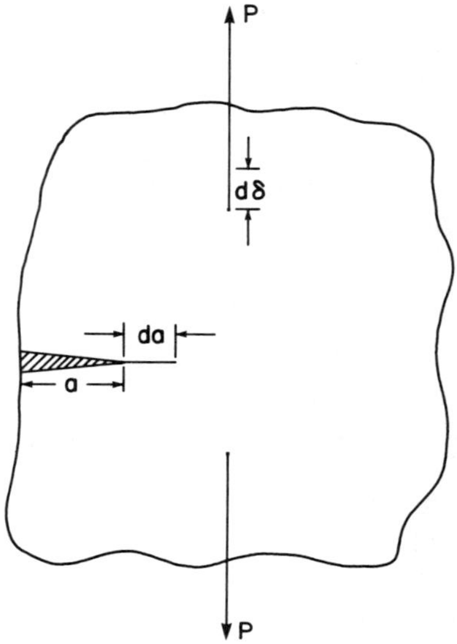

Fig. 4-68—Cracked plate subjected to tensile load

$$G_I = -\frac{1}{h}\frac{dU}{da} \qquad (4.134)$$

where U is the total strain energy stored in the test specimen and h is the plate thickness. From Fig. 4-69 it can be seen that the strain energy lost due to the crack extending from a to $a + da$ for a linear elastic body is simply the area, dA, between the loading and unloading curves. Thus,

$$-dU = dA = \tfrac{1}{2}(Pd\delta - \delta dP) \qquad (4.135)$$

where P and δ are the load and resulting deflection, respectively, as illustrated in Figs. 4-68 and 4-69. Combining eqs (4.134) and (4.135), we obtain the expression

$$G_I = \frac{1}{h}\frac{dA}{da} = \frac{1}{2h}\left(P\frac{d\delta}{da} - \delta\frac{dP}{da}\right) \qquad (4.136)$$

4.5.2 The Double Cantilever Beam Test

The DCB test was originally developed for the purpose of evaluating Mode I fracture of adhesive bonded joints. Both a tapered and a

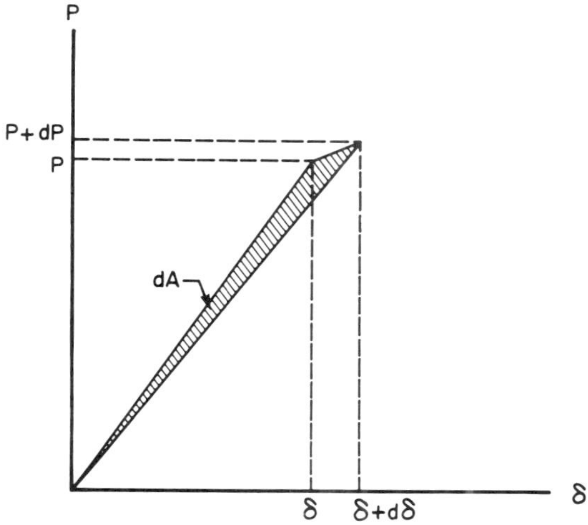

Fig. 4-69—Strain energy release rate concept (Whitney and Browning[69])

straight-sided DCB specimen have been utilized in bonded joint characterization. Bascom[67] applied a tapered DCB specimen to the evaluation of woven fabric composites. Wilkins, *et al.*[68] utilized a straight-sided DCB specimen in conjunction with 0 deg unidirectional and $[0/90]_s$ laminates constructed from graphite-epoxy prepreg. The straight-sided DCB test specimen has become considerably more popular than the tapered DCB specimen.

A DCB test specimen is illustrated in Fig. 4-70. Typical dimensions are L = 230 mm (9 in.) with b = 25 mm (1 in.). The specimen contains a starter crack, a_0, usually in the form of Teflon film with a_0 = 25 mm

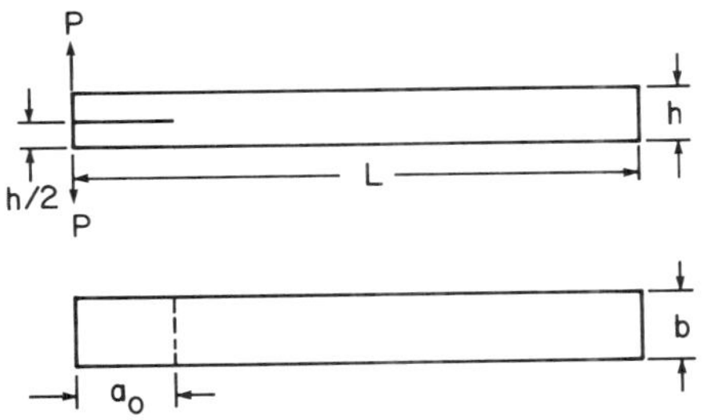

Fig. 4-70—Double cantilever beam specimen

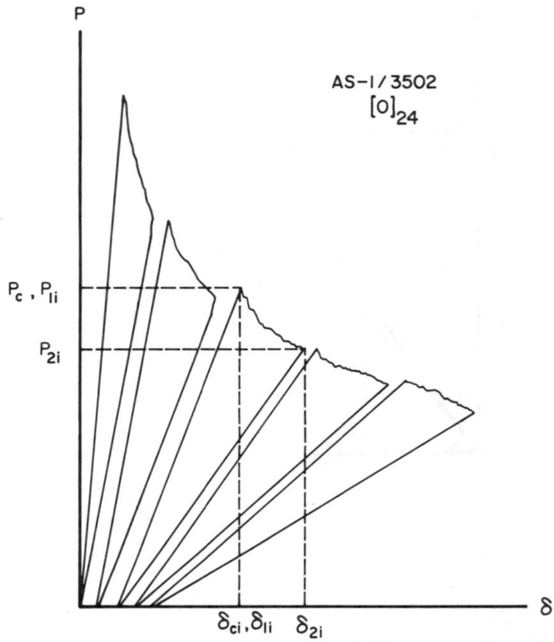

Fig. 4-71—Loading and unloading curves for typical double cantilever beam specimen (Whitney and Browning[69])

(1 in.), placed along the center line. Specimen thickness should be at least 2.5 mm (0.1 in.). End tabs must be arranged such that the load P remains vertical during the test. This can be accomplished by utilizing a piano-hinge type tab[68] or an extruded aluminum "T" type tab.[69] In the latter case, the tabs contain holes which attach the specimen to the loading fixture, and slots for adjustable screws which attach to an extensometer for measuring crack-opening displacement. For the piano-hinge type end tab, crack-opening displacement can be determined from cross-head travel. Obviously more accuracy can be obtained by using an extensometer.

A cross-head speed in the range of 1.27 mm/min (0.05 in./min) is recommended. For displacement of controlled tests, crack growth in this specimen is stable. Crack length can be tracked by marking the specimen edges at desired increments with a silver-leaded pencil. There is usually a pop-in phenomenon associated with the first increment of crack extension which results in very erratic loading and unloading behavior. Thus, this first increment of crack propagation should be ignored. In addition, this first increment of crack growth produces a

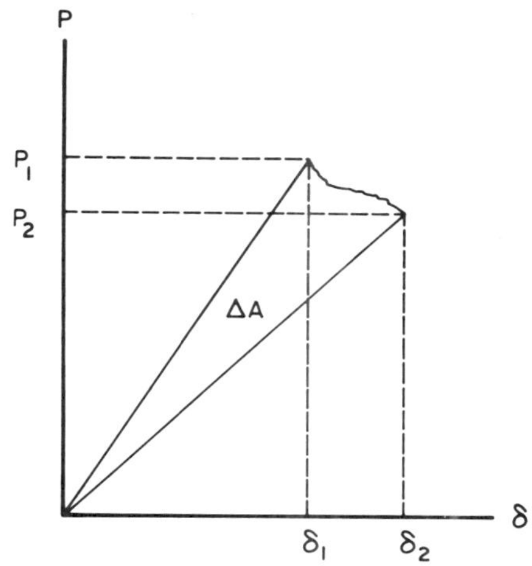

Fig. 4-72—The area method for determining G_{IC} (Whitney and Browning[69])

natural crack from which G_{IC} can be determined on further increments of crack propagation. Depending on the data-reduction scheme, further data can be obtained by a series of loading and unloading curves (see Fig. 4-71) at desired crack-extension intervals Δa, or by noting crack length on a continuous plot of load vs. deflection during stable crack growth. A number of data-reduction schemes can be utilized in conjunction with the DCB test specimen.

The compliance method as developed in eq (4.135) is also applicable to the case of finite crack extension and provides the basis for a straightforward data-reduction scheme. Consider a typical loading and unloading curve for a crack propagating in a unidirectional graphite/epoxy composite, as shown in Fig. 4-72. The DCB specimen is loaded linearly to P_1 where the crack begins to extend. During crack extension from a to $a + \Delta a$, the load drops to P_2. If the specimen is then unloaded, the loss in strain energy due to crack extension is simply the area, ΔA, between the loading and unloading curves. For cases in which the load deflection curve during crack propagation can be approximated by a straight line, one can determine the critical strain energy release rate, G_{IC}, from the relationship

$$G_{IC} = \frac{1}{b}\,\frac{\Delta A}{\Delta a} = \frac{1}{2b\Delta a} = (P_1\delta_2 - P_2\delta_1) \qquad (4.137)$$

An average value of G_{IC} can be determined by measuring P_1, P_2, δ_1, and δ_2 for a series of N crack extensions of length Δa. Thus,

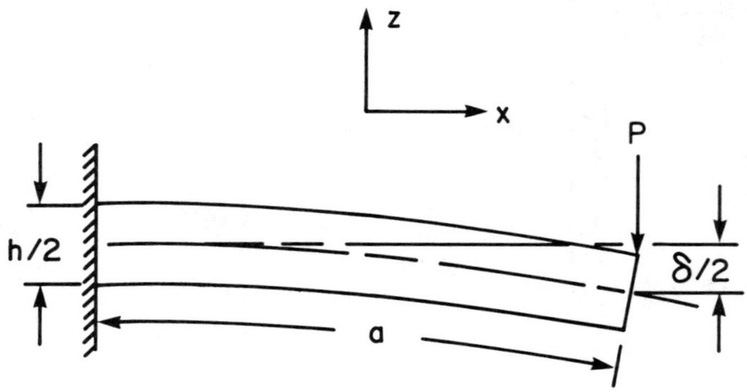

Fig. 4-73—Cantilever beam analysis model for determining G_{IC}

$$G_{IC} = \frac{1}{2bN\Delta a} \sum_{i=1}^{N} (P_{1i}\delta_{2i} - P_{2i}\delta_{1i}) \tag{4.138}$$

In cases where the loading and/or unloading curves for elastic beams are nonlinear, the strain energy release rate can be determined by measuring the area between the loading and unloading curves. Large deflections or nonlinear elastic stress–strain behavior provide sources for nonlinear load-deflection response.

One of the most common approaches to analyzing a DCB specimen is to assume that each cracked half behaves as the cantilever beam illustrated in Fig. 4-73. Under such an assumption

$$\delta = BPa^3 \tag{4.139}$$

where B is a constant defined by

$$B = \frac{64}{E_x^b bh^3} \tag{4.140}$$

and E_x^b is the effective bending modulus in the axial direction of a cantilever beam of thickness $h/2$, as illustrated in Fig. 4-73. Substituting eq (4.139) into eq (4.136), one obtains the energy release rate relationship

$$G_I = \frac{3BP^2a^2}{2b} = \frac{3P\delta}{2ba} \tag{4.141}$$

As a matter of historical interest, if one uses eq (4.139) in conjunction with eq (4.136) for either the case of fixed grip (δ = constant) or the case of fixed load (P = constant), eq (4.141) is recovered. Thus, the cases of fixed grip and fixed load yield identical results. This is important because other authors have obtained expressions for the fixed load and fixed grip case in which G_I are opposite in sign.

The critical strain energy release rate is defined from eq (4.141)

$$G_{IC} = \frac{3BP_c^2a^2}{2b} \tag{4.142}$$

or

$$G_{IC} = \frac{3P_c\delta_c}{2ba} \tag{4.143}$$

where P_c and δ_c are critical values of P and δ, respectively, associated with the onset of crack growth. In an actual experiment, continuous loading and crack extension can be simultaneously induced. In such a case the energy release rate is equal to G_{IC} during loading, after the onset of crack extension, and eq (4.143) can be directly applied at a number of crack lengths, *i.e.*,

$$G_{IC} = \frac{3H}{2b} \tag{4.144}$$

where

$$H = \frac{P_c\delta_c}{a} = \text{constant} \tag{4.145}$$

A continuous plot of P_c vs. δ_c during crack extension can then be used to measure H by directly applying eqs (4.144) and (4.145) at a number of crack lengths and then calculating an average value, *i.e.*,

$$H = \frac{1}{N}\sum_{i=1}^{N}\frac{P_{ci}\delta_{ci}}{a_i} \tag{4.146}$$

where P_{ci} and δ_{ci} are the critical values of P and δ associated with the ith crack length a_i, and N is the total number of data points.

Whenever the beam analysis is used in conjunction with the determination of G_{IC}, one must be aware of the effect of large deflections. This has been discussed in detail by Devitt, Schapery, and Bradley.[70] If large deflections are present, then G_{IC} must be determined from a nonlinear beam analysis[70] in conjunction with eq (4.136), *i.e.*, eqs (4.142) and (4.143) are no longer valid.

Another data-reduction scheme involves an empirical generalization of eq (4.139) proposed by Berry.[71] In particular, he assumed

$$\delta = RPa^n \tag{4.147}$$

where R and n are constants determined experimentally from the relationship

$$\log(P/\delta) = -\log R - n \log a \tag{4.148}$$

A least squares fit to eq (4.148) for a series of loading and unloading curves, as shown in Fig. 4-71, allows R and n to be determined. Obviously beam theory is recovered if $n = 3$.

The critical strain energy release rate is determined by substituting eq (4.147) into eq (4.136) with the result

$$G_{IC} = \frac{nRP_c^2 a^{n-1}}{2b} \tag{4.149}$$

or

$$G_{IC} = \frac{nP_c\delta_c}{2ba} \tag{4.150}$$

Using eq (4.150) in conjunction with eq (4.146), one obtains

$$G_{IC} = \frac{nH}{2b} \tag{4.151}$$

Comparing eqs (4.151) and (4.144), one sees that the difference between the empirical analysis and the beam analysis depends on the value of n, *i.e.*,

$$\frac{G_{IC}^e}{G_{IC}^b} = \frac{n}{3} \tag{4.152}$$

where subscripts e and b denote the empirical analysis and beam analysis, respectively. It should be noted that eq (4.147) leads to very unusual units for the effective compliance term, R.

Viscoelastic behavior in conjunction with the DCB specimen has been discussed by Devitt, Schapery, and Bradley.[70] It was concluded that an elastic strain energy release rate calculation could be utilized for viscoelastic materials provided the axial stress–strain relation is essentially elastic and viscoelastic effects are limited to a small area around the crack tip. This is generally the case for high-modulus filament-dominated laminates under room temperature conditions. For cases where significant viscoelastic response is observed, the area between the loading and unloading curves depends on crack speed, load/deflection rate, etc.

The data in Fig. 4-71 were obtained from Ref. 69. Values of G_{IC} were obtained from this data by using the three data-reduction schemes discussed. A comparison of results is shown in Table 4-4. As noted in Fig. 4-71, these data are obtained in conjunction with 0-deg unidirectional AS-1/3502 graphite/epoxy composites containing 24 plies. Additional DCB data from Ref. 69 are displayed in Table 4-5 for various 0-deg unidirectional graphite fiber reinforced composites. All of the data were reduced by the area method with a_2 indicating the crack length at the time of unloading. Values of G_{IC} are shown for each 0.5-in. increment of crack extension in order to illustrate typical data scatter with change in crack length.

Interpretation of any data obtained on the DCB test should be accomplished only after careful examination of the failure surface. The matrix dominated type of fracture surface always represents a minimum value of G_{IC}. Thus, for advanced composites unidirectional data should be emphasized. In the case of multidirectional laminates, the crack tends to jump and move along ply interfaces above and below the laminate midplane, creating nonself-similar crack growth which often

Table 4-4(a) Data from Fig. 4-71

AS-1/3502 $[0]_{24}$				
a (in.)	P_c, P_1 (lb)	δ_c, δ_1 (in.)	P_2 (lb)	δ_2 (in.)
2.0	14.3	0.058	11.1	0.091
2.5	10.9	0.093	8.8	0.150
3.0	8.9	0.155	7.2	0.230
3.5	7.3	0.235	6.3	0.317
4.0	6.4	0.335	5.5	0.411

Table 4-4(b) Comparison of Methods for Determining G_{IC} (Ref. 69)

AS-1/3502 Unidirectional	
Method	G_{IC} (lb/in.)*
DCB, $[0]_{24}$, Area Method, eq (5)	0.801
DCB $[0]_{24}$, Beam Analysis, eq (11)	0.691
DCB, $[0]_{24}$, Empirical Analysis, eq (22)	0.861

**For SI units 1 lb/in. = 0.175 kJ/m²*

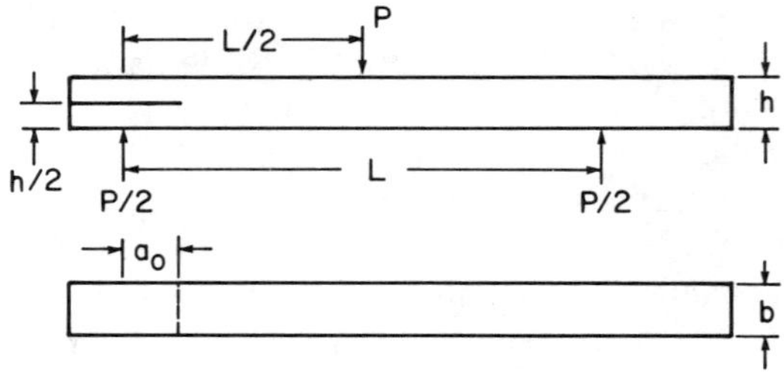

Fig. 4-74—End notch flexure test specimen

yields apparent values of G_{IC} considerably higher than those obtained with a matrix dominated by 0-deg unidirectional specimen. Such behavior has been documented by Wilkins *et al.*[68] and by Nicholls and Gallagher.[72] Even in the case of the unidirectional specimen, the crack can move slightly off the center line creating "fiber bridging," which also yields higher values of G_{IC} than the pure matrix fracture type of delamination.

The DCB test can also be utilized for measuring G_{IC} on neat resins. In this case the specimen is prepared in the same manner as in a classical adhesive bonded joint test with the resin bonded between two aluminum adherends.

4.5.3 *The End Notch Flexure Test*

The ENF test was developed by Russell and Street.[73] This method is based on a three-point flexure test as illustrated in Fig. 4-74. Recom-

Table 4-5 DCB Energy Release Rates as Determined by Area Method (Ref. 69)

	a_2 (in.)					
	2.0	2.5	3.0	3.5	4.0	
Material	G_{IC} (lb/in.)*					Average G_{IC} (lb/in.)
AS-1/3502 $[0]_{24}$	0.658	0.817	0.931	0.834	0.767	0.801
AS-4/3502 $[0]_{24}$	0.814	0.746	1.097	0.950	0.986	0.919
T300/V378A $[0]_{24}$	0.480	0.418	0.366	0.545	0.255	0.413
AS-1/Polysulfone $[0]_{12}$	3.51	2.97	3.82	2.97	3.41	3.34

**For SI units 1 lb/in. = 0.175 kJ/m²*

mended specimen dimensions are $L = 100$ mm (4 in.) with $b = 20$ mm (0.75 in.). As in the DCB specimen, a midplane starter crack of a desired length a_0 is introduced at one end of the beam. Teflon film is a suitable material for the starter crack. The specimen thickness should be chosen such that a desired span-to-depth ratio, L/h, is obtained. As a general guideline, L/h values can be chosen which are compatible with a conventional three-point flexure test as covered by ASTM Standard D-790. For graphite/epoxy materials, $L/h = 32$ is recommended. Loading rates consistent with ASTM D-790 should also be chosen. Crack propagation in the ENF specimen tends to be unstable. As a result Russell and Street[73] were able to obtain only two fracture energy values, G_{IIC}, for each specimen tested. The first value was associated with crack growth from the root of the starter notch and the second value associated with extension of the arrested initial crack which averaged approximately 25 mm (1 in.).

The unstable nature of the crack growth in the ENF test renders the area data-reduction method much less useful than in the DCB test. Russell and Street[73] used the beam illustrated in Fig. 4-75 to derive a compliance vs. crack length relationship from classical beam theory with the result

$$\delta = \frac{(L^3 + 12a^3)P}{4E_x^b bh^3} \tag{4.153}$$

where δ is the beam deflection under the load nose. Substituting eq (4.153) into eq (4.136), one obtains the Mode II critical energy release rate

$$G_{IIC} = \frac{9a^2P_c^2}{2E_x^b b^2 h^3} \tag{4.154}$$

or

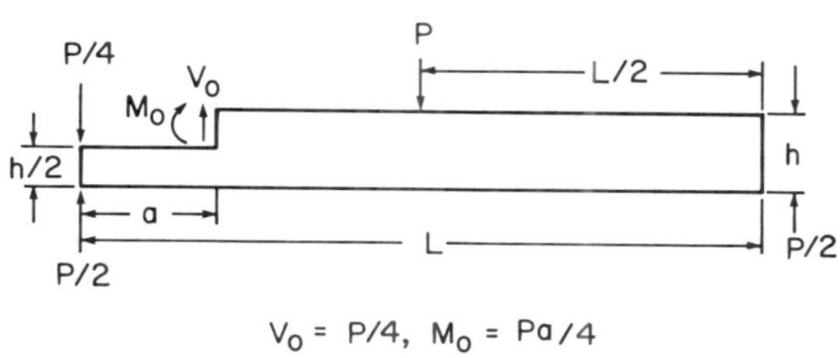

Fig. 4-75—Beam model for analyzing end notch flexure specimen

Table 4-6 ENF Energy Release Rates (Ref. 73)

Material $[0]_{24}$	T (° F)	G_{IIC} (lb/in.)*	G'_{IIC} (lb/in.)
AS1/3501-6	68	2.58	2.20
AS1/3501-6	−58	3.18	2.20
AS1/3501-6	212	2.31	2.10
HMS/3501-6	68	0.868	0.760

**For SI units 1 lb/in. = 0.175 kJ/m²*

$$G_{IIC} = \frac{18a^2 P_c \delta_c}{(L^3 + 12a^3)} \tag{4.155}$$

where, as previously, P_c and δ_c are critical values of P and δ, respectively, associated with the onset of crack growth, E_x^b is the effective bending modulus of the beam, and G_{IIC} is the Mode II critical energy release rate. The deflection δ can be measured by cross-head travel or with a deflectometer as outlined in ASTM D-790.

Experimental results obtained by Russell and Street[73] are shown in Table 4-6 for 0-deg unidirectional graphite/epoxy composites at various temperatures dry. The values of G_{IIC} are determined from initial crack growth from the root of the starter crack, while G_{IIC} corresponds to extension from the arrested initial crack. It should be noted that the root of the arrested initial crack is almost directly under the load nose. This may cause a large normal compression stress on the root of the crack which is not present in the original starter crack.

A number of angle-ply laminates were also tested by Russell and Street.[73] As in the case of the DCB specimen, the crack deviated from the midplane creating a mixed mode type of fracture. Thus, it is recommended that the ENF test be limited to 0-deg unidirectional composites.

REFERENCES

1. Hahn, H.T. and Kim, R.Y., "Swelling of Composite Laminates," *Environmental Effects on Composite Materials,* ASTM STP 658, American Society for Testing and Materials, Philadelphia, 98-120 (1978).
2. Hahn, H.T., "Residual Stresses in Polymer Matrix Composite Laminates," *Journal of Composite Materials,* Vol. 10, No. 4, 266-278 (Oct. 1976).

3. Whitney, J.M., Browning, C.E. and Mair, A., "Analysis of the Flexure Test for Laminated Composite Materials," *Composite Materials: Testing and Design (Third Conference),* ASTM STP 546, American Society for Testing and Materials, 30-45 (1974).
4. Weibull, W., "A Statistical Theory of the Strength of Materials," *Ing. Vetenskaps Akad, Handl.,* No. 151, 5-45 (1939).
5. Halpin, J.C., Jerina, K.L. and Johnson, T.A., "Characterization of Composites for the Purpose of Reliability Evaluation," *Analysis of the Test Methods for High Modulus Fibers and Composites,* ASTM STP 521, American Society for Testing and Materials, Philadelphia, 5-64 (1973).
6. Kaminski, B.E., "Effect of Specimen Geometry on the Strength of Composite Materials," *Analysis of the Test Methods for High Modulus Fibers nd Composites,* ASTM STP 521, American Society for Testing and Materials, Philadelphia, 181-191 (1973).
7. Hahn, H.T. and Kim, R.Y., "Proof Testing of Composite Materials," *Journal of Composite Materials,* Vol. 9, 297-311 (July 1975).
8. Weil, N.A. and Daniel, I.M., "Analysis of Fracture Probabilities in Nonuniformly Stressed Brittle Materials," *Journal of the American Ceramic Society,* Vol. 47, 268-274 (June 1964).
9. Pagano, N.J. and Halpin, J.C., "Influence of End Constraint in the Testing of Anisotropic Bodies," *Journal of Composite Materials,* Vol. 2, No. 1, 18-31 (Jan. 1968).
10. Hofer, K.E., Jr., Rao, N. and Larsen, D., "Development of Engineering Data on Mechanical Properties of Advanced Composite Materials," Air Force Technical Report *AFML-TR-72-205, Part I* (Sept. 1972).
11. Verette, R.M. and Labor, J.D., "Structural Criteria for Advanced Composites," Air Force Technical Report *AFFDL-TR-76-142, Vol. 1, Summary* (March 1977).
12. Kasen, M.B., Schramm, R.E. and Read, D.T., "Fatigue of Composites at Cryogenic Temperatures," *Fatigue of Filamentary Composites,* ASTM STP 636, K.L. Reifsnider and K.N. Lauraitis, Eds., American Society for Testing and Materials, 141-151 (1977).
13. Grimes, G.C., Francis, P.H., Commerford, G.E. and Wolfe, G.K., "An Experimental Investigation of the Stress Levels at Which Significant Damage Occurs in Graphite Fiber Plastic Composites," Air Force Technical Report, *AFML-TR-72-40* (May 1972).
14. Ryder, J.T. and Black, E.D., "Compression Testing of Large Gage Length Composite Coupons," *Composite Materials: Testing and Design (Fourth Conference),* ASTM STP 617, American Society for Testing and Materials, Philadelphia, 170-189 (1977).
15. Ryder, J.T. and Walter, E.K., "Effect of Compression on Fatigue Properties of a Quasi-Isotropic Graphite/Epoxy Composite," *Fatigue of Filamentary Composite Materials,* ASTM STP 636, K.L. Reifsnider and K.N. Lauraitis, Editors, American Society for Testing and Materials, Philadelphia, 3-26 (1977).
16. Lantz, R.B., "Boron Epoxy Laminate Test Methods," *Journal of Composite Materials,* Vol. 3, No. 4, 642-650 (Oct. 1969).

17. "Uniaxial Compression," Section 4.2.4, *Advanced Composites Design Guide,* Vol. 4, Materials, 3rd Edition, Air Force Flight Dynamics Laboratory, (Dec. 1975).
18. Lantz, R.B. and Baldridge, K.G., "Angle-Plied Boron/Epoxy Test Method—A Comparison of Beam-Tension and Axial Tension Coupon Testing," *Composite Materials: Testing and Design,* ASTM STP 460, American Society for Testing and Materials, Philadelphia, 94-107 (1969).
19. Rosen, B.W., "A Simple Procedure for Experimental Determination of the Longitudinal Shear Modulus of Unidirectional Composites," *Journal of Composite Materials,* Vol. 6, No. 4, 552-554 (Oct. 1972).
20. Chamis, C.C. and Sinclair, J.H., "Ten-degree Off-axis Test for Shear Properties in Fiber Composites," *Experimental Mechanics,* Vol. 17, No. 9, 339-346 (Sept. 1977).
21. Daniel, I.M., "Biaxial Testing of Graphite/Epoxy Composites Containing Stress Concentrations," Air Force Technical Report *AFML-TR-76-244,* Part I (Dec. 1976).
22. Whitney, J.M., Stansbarger, D.L. and Howell, H.B., "Analysis of the Rail Shear Test—Applications and Limitations," *Journal of Composite Materials,* Vol. 5, No. 1, 24-35 (Jan. 1971).
23. Adams, D.F. and Thomas, R.L., "Test Methods for the Determination of Unidirectional Composite Shear Properties," *Advances in Structural Composites,* Proceedings 12th SAMPE Symposium (Paper AC-5) (1967).
24. Sullivan, T.L. and Chamis, C.C., "Some Important Aspects in Testing High-Modulus Fiber Composite Tubes in Axial Tension," *Analysis of the Test Methods for High Modulus Fibers and Composites,* ASTM STP 521, American Society for Testing and Materials, 277-292 (1973).
25. Hahn, H.T. and Erikson, J., "Characterization of Composite Laminates Using Tubular Specimens," Air Force Technical Report *AFML-TR-77-144,* (Aug. 1977).
26. Whitney, J.M. and Halpin, J.C., "Analysis of Laminated Tubes Under Combined Loading," *Journal of Composite Materials,* Vol. 2, No. 3, 360-367 (July 1968).
27. Pagano, N.J. and Whitney, J.M., "Geometric Design of Composite Cylindrical Characterization Specimens," *Journal of Composite Materials,* Vol. 4, No. 3, 360-378 (July 1970).
28. Pagano, N.J., "Stress Gradients in Laminated Composite Cylinders," *Journal of Composite Materials,* Vol. 5, No. 2, 260-265 (April 1971).
29. Whitney, J.M., "On the Use of Shell Theory for Determining Stresses in Composite Cylinders," *Journal of Composite Materials,* Vol. 5, No. 3, 340-353 (July 1971).
30. Grimes, G.C. and Francis, P.H., "Investigation of Stress Levels Causing Significant Damage," Air Force Technical Report *AFML-TR-75-33* (June 1975).
31. Pipes, R.B., Reed, D.L. and Ashton, J.E., "Experimental Determination of Interlaminar Shear Properties of Composite Materials," *SESA Paper No. 1985 A,* Presented at 1972 SESA Spring Meeting, Cleveland, OH, May 23-26, 1972.

32. Berg, C.A., Tirosh, J. and Israeli, M., "Analysis of Short Beam Bending of Fiber Reinforced Composites," *Composite Materials: Testing and Design (Second Conference),* ASTM STP 497, American Society for Testing and Materials, 206-218 (1972).
33. Konish, H.J., Jr., Swedlow, J.L. and Cruse, T.A., "Experimental Investigation of Fracture in an Advanced Fiber Composite," *Journal of Composite Materials,* Vol. 6, No. 1, 114-125 (Jan. 1972).
34. Mandell, John F., Wang, Su-Su and McGarry, Frederick, "The Extension of Crack Tip Damage Zones in Fiber Reinforced Plastic Laminates," *Journal of Composite Materials,* Vol. 9, No. 3, 266-287 (July 1975).
35. Nuismer, R.J. and Whitney, J.M., "Uniaxial Failure of Composite Laminates Containing Stress Concentrations," *Fracture Mechanics of Composites,* ASTM STP 593, American Society for Testing and Materials, 117-142 (1975).
36. Paris, Paul C. and Sih, George C., "Stress Analysis of Cracks," *Fracture Toughness Testing,* ASTM STP 381, American Society for Testing and Materials, 30-83 (1965).
37. Brown, William F., Jr. and Srawley, John E., *Plane Strain Crack Toughness Testing of High Strength Metallic Materials,* ASTM STP 410, American Society for Testing and Materials, 11 (1969).
38. Snyder, M.D. and Cruse, T.A., "Crack Tip Stress Intensity Factors in Finite Anisotropic Plates," Air Force Materials Laboratory Report *AFML-TR-73-209* (Aug. 1973).
39. Morris, D.H. and Hahn, H.T., "Mixed-Mode Fracture of Graphite/Epoxy Composites: Fracture Strength," *Journal of Composite Materials,* Vol. 11, No. 2, 124-138 (April 1977).
40. Toda, H., Paris, P.C. and Irwin, G.R., *The Stress Analysis of Cracks Handbook,* Del Research Corporation, Hellertown, PA, 199 (1973).
41. Peterson, R.E., *Stress Concentration Factors,* Wiley-Interscience, John Wiley and Sons, New York, NY, 110-111 (1974).
42. Daniel, I.M., "Biaxial Testing of Graphite/Epoxy Composites Containing Stress Concentrations," Air Force Materials Laboratory Report *AFML-TR-76-244, Part II* (June 1977).
43. Waddoups, M.E., Eisenmann, J.R. and Kaminski, B.E., "Macroscopic Fracture Mechanics of Advanced Composite Materials," *Journal of Composite Materials,* Vol. 5, No. 4, 446-454 (Oct. 1971).
44. Whitney, J.M., Browning, C.E. and Grimes, G.C., "The Relationship Between Significant Damage and Stress–Strain Response of Laminated Polymeric Matrix Composites," *Composite Materials in Engineering Design,* Edited by Bryan R. Noton, American Society for Metals, Metals Park, OH, 441-447 (1973).
45. Reifsnider, K.L., Henneke, E.G. II and Stinchcomb, W.W., "Delamination in Quasi-Isotropic Graphite-Epoxy Laminates," *Composite Materials: Testing and Design (Fourth Conference),* ASTM STP 617, American Society for Testing and Materials, 93-105 (1977).
46. Pagano, N.J. and Hahn, H.T., "Evaluation of Composite Curing Stresses," *Composite Materials: Testing and Design (Fourth Conference),* ASTM STP 617, American Society for Testing and Materials, 317-329 (1977).

47. Petit, P.H. and Waddoups, M.E., "A Method of Predicting the Nonlinear Behavior of Laminated Composites," *Journal of Composite Materials,* Vol. 3, No. 1, 2-19 (Jan. 1969).
48. Whitney, J.M., Pagano, N.J. and Pipes, R.B., "Design and Fabrication of Tubular Specimens for Composite Characterization," *Composite Materials: Testing and Design (Second Conference),* ASTM STP 497, American Society for Testing and Materials, Philadelphia, 52-67 (1972).
49. Cole, B.W. and Pipes, R.B., "Utilization of the Tubular and Off-Axis Specimens for Composite Biaxial Characterization," *Proceedings of the Conference on Fibrous Composites in Flight Vehicle Design,* Air Force Technical Report AFFDL-TR-72-130, 973-1020 (Sept. 1972).
50. Whitney, J.M., Grimes, G.C. and Francis, P.H., "Effect of End Attachment on the Strength of Fiber-Reinforced Composite Cylinders," *Experimental Mechanics,* Vol. 13, No. 5, 185-192 (May 1973).
51. Nagy, A. and Lindholm, U.S., "Hydraulic Grip System for Composite Tube Specimens," Air Force Technical Report *AFML-TR-73-239* (Nov. 1973).
52. Pagano, N.J., Halpin, J.C. and Whitney, J.M., "Tension Buckling of Anisotropic Cylinders," *Journal of Composite Materials,* Vol. 2, No. 2, 154-167 (April 1968).
53. Whitney, J.M. and Sun, C.T., "A Refined Theory for Laminated Anisotropic Cylindrical Shells," *Journal of Applied Mechanics,* Vol. 41, No. 2, 471-476 (June 1974).
54. Vicario, A.A. and Rizzo, R.R., "Effect of Length on Laminated Thin Tubes Under Combined Loading," *Journal of Composite Materials,* Vol. 4, No. 2, 273-277 (April 1970).
55. Yang, J.N. and Liu, M.D., "Residual Strength Degradation Model and Theory of Periodic Proof Tests for Graphite-epoxy Laminates," *Journal of Composite Materials,* Vol. 11, No. 3, 176-203 (April 1977).
56. Hahn, H.T. and Kim, R.Y., "Fatigue Behavior of Composite Laminate," *Journal of Composite Materials,* Vol. 10, No. 2, 156-180 (April 1976).
57. Mann, N.R., Schafer, R.E. and Singpurwalla, N.D., *Methods for Statistical Analysis of Reliability and Life Data,* John Wiley and Sons, New York (1974).
58. Thoman, Darrell R. and Bain, Lee J., "Two Sample Tests in the Weibull Distribution," *Technometrics,* Vol. 11, No. 4, 505-515 (Nov. 1969).
59. Wolff, R.V. and Lemon, G.H., "Reliability Prediction for Composite Joints-Bonded and Bolted," Air Force Technical Report *AFML-TR-74-197,* 66-74 (March 1976).
60. Thoman, Darrell R., Bain, Lee J. and Antle, Charles E., "Inferences on the Parameters of the Weibull Distribution," *Technometrics,* Vol. 11, No. 3, 445-460 (Aug. 1969).
61. Cohen, Clifford A., "Maximum Likelihood Estimation in the Weibull Distribution Based on Complete and Censored Samples," *Technometrics,* Vol. 7, No. 4, 579-588 (Nov. 1965).
62. Pipes, R.B., Kaminski, B.E. and Pagano, N.J., "Influence of the Free Edge Upon the Strength of Angle-Ply Laminates," *Analysis of the Test Methods for High Modulus Fibers and Composites,* ASTM STP 521, American Society for Testing and Materials, Philadelphia, 181-191 (1973).

63. Pipes, R.B. and Cole, B.W., "On the Off-Axis Strength Test for Anisotropic Materials," *Journal of Composite Materials,* Vol. 7, No. 2, 246-256 (April 1973).
64. Whitney, J.M. and Browning, C.E., "On Interlaminar Beam Experiments for Composite Materials," *Proceedings of the V International Congress on Experimental Mechanics,* SEM, 97-101 (June 1984).
65. Browning, C.E., Abrams, F.L. and Whitney, J.M., "A Four-Point Shear Test for Graphite/Epoxy Composites," *Composite Materials: Quality Assurance and Processing,* ASTM STP 797, ed. C.E. Browning, ASTM, 54-74 (1983).
66. Whitney, J.M., "Elasticity Analysis of Orthotropic Beams Under Concentrated Loads," *Composite Science and Technology,* Vol. 1 (1985).
67. Bascom, W.D., Bitner, R.J., Moulton, R.J. and Siebert, A.R., "The Interlaminar Fracture of Organic-Matrix Woven Reinforced Composites," *Composites,* 9-14 (Jan. 1980).
68. Wilkins, D.J., Eisenmann, Camin, R.A. and Margolis, W.S., "Characterizing Delamination Growth in Graphite-Epoxy," *Damage in Composite Materials: Basic Mechanisms, Accumulation, Tolerance and Characterization,* ASTM STP 775, ed. K.L. Reifsnider, ASTM, 168-183 (1982).
69. Whitney, J.M. and Browning, C.E., "A Double Cantilever Beam Test for Characterizing Mode I Delamination of Composite Materials," *Journal of Reinforced Plastics and Composites,* Vol. 1, 297-313 (Oct. 1982).
70. Devitt, D.F., Schapery, R.A. and Bradley, W.L., "A Method for Determining Mode I Delamination Fracture Toughness of Elastic and Viscoelastic Composite Materials," *Journal of Composite Materials,* Vol. 14, 270-285 (Oct. 1980).
71. Berry, J.P., "Determination of Fracture Surface Energies by the Cleavage Technique," *Journal of Applied Physics,* Vol. 34, 62-66 (1983).
72. Nicholls, D.J. and Gallagher, "Determination of G_{IC} in Angle-Ply Composites Using a Cantilever Beam Test Method," *Journal of Reinforced Plastics and Composites,* Vol. 2, 2-17 (Jan. 1983).
73. Russell, A.J. and Street, K.N., "Factors Affecting the Interlaminar Fracture Energy of Graphite/Epoxy Laminates," *Proceedings of the Fourth International Conference on Composite Materials,* Elsevier, North Holland, 279-286 (1984).

5

EFFECTS OF MOISTURE

5.1 GLASS TRANSITION TEMPERATURE

The glass transition temperature, T_g, of a polymer is defined as the temperature above which it is soft and below which it is hard. The hard polymer is a glasslike material, while the soft polymer varies from a rubbery material to an oil.[1] For cross-linked amorphous polymers, such as epoxy resins, T_g is determined by the cure or post cure temperature. Thus, T_g can be increased for these systems by raising the cure or post cure temperature.

5.1.1 Viscoelastic Properties

From a practical standpoint it is more appropriate to discuss a glass transition temperature region rather than a single glass transition temperature, as the change from a hard polymeric material to a soft material takes place over a temperature range. This range, as illustrated in Fig. 5-1 for a typical cross-linked amorphous polymer, is characterized by a decreasing modulus with increasing temperature. A very rapid decrease in modulus occurs with increasing temperatures above

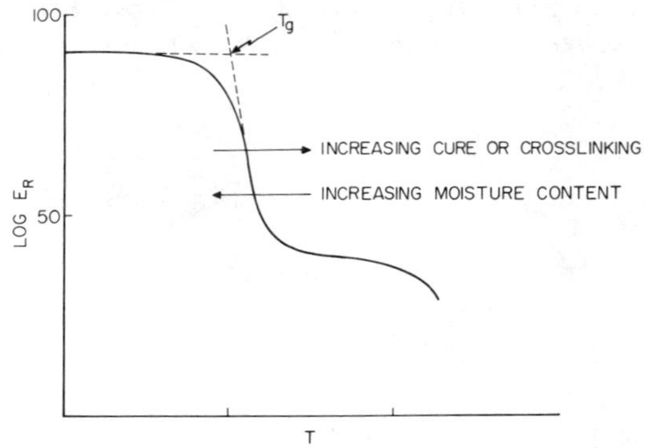

Fig. 5-1—Modulus, E_R, versus temperature for typical epoxy resin illustrating glass transition temperature (Browning, Husman, and Whitney[5])

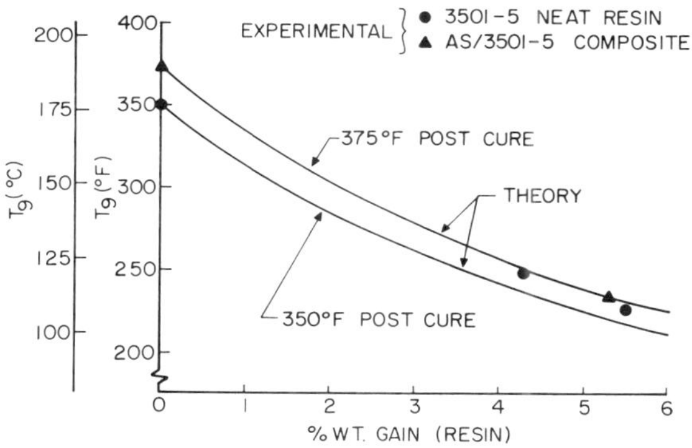

Fig. 5-2—Glass transition temperature as a function of absorbed moisture in resin (Browning, Husman, and Whitney[5])

T_g. Thus, viscoelastic properties are a concern over a range of temperatures below as well as above T_g.

5.1.2 Effect of Moisture on Glass Temperature

It is well recognized[1] that the T_g of a polymer can be lowered by mixing with it a miscible liquid (diluent) that has a lower glass transition temperature than the polymer. This process is referred to as plasticization. Many polymers, such as epoxy resins, absorb moisture from high humidity environment.[2-4] Thus, moisture is a plasticizer for such systems, producing a lower value of T_g. Similar effects are observed in composites.[5]

The lowering of T_g with increasing moisture content is illustrated in Fig. 5-2 for both an epoxy resin and derived graphite/epoxy composite. Theoretical results are based on the Bueche-Kelley theory for T_g of a polymer-diluent system.[6] It should be noted that the dry T_g is higher for the composite than for the resin. This is due to the fact that the resin was not postcured, while the composite was postcured at 191° C (375° F).

5.1.3 Measurement of T_g

A number of methods are available for measuring T_g of polymers. The procedure which appears to give the most consistent results and is easily adaptable for both resins and composites is the heat distortion temperature (HDT) method. This test consists of the three-point loading of a flexure specimen surrounded by an increasing temperature environment. A fixed load is applied at the center of the beam and the deflection

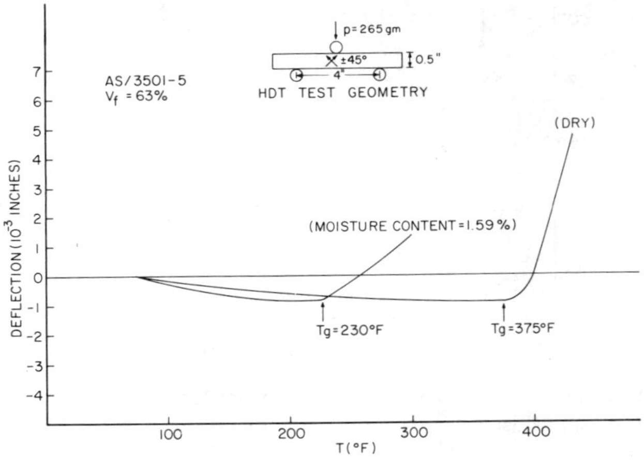

Fig. 5-3—Typical HDT test results for AS/3501-5 graphite/epoxy composite $(90/\pm45)_s$ (Browning, Husman, and Whitney[5])

measured as a function of temperature. The glass transition region is characterized by a rapid increase in the coefficient of thermal expansion. The choice of actual T_g is somewhat arbitrary; however, the point where the slope of the temperature–deflection curve becomes positive yields very consistent results. The test procedure for polymeric materials is described in ASTM Standard D648-72.

For composite materials the test procedure is essentially the same as for polymeric specimens. Typical results, test geometry, and applied load are shown in Fig. 5-3 for a matrix dominated laminate specimen having the orientation $[90/\pm45]_s$. The specimen is placed on edge so that the bending deformation takes place inplane rather than through-the-thickness. An oil bath surrounds the specimen as a heat-transfer medium and also decelerates the drying process for wet T_g measurements.

5.2 MOISTURE DIFFUSION

Because of the effect of moisture in lowering of the resin's glass transition temperature, there is considerable interest in determining the mechanical properties of composite laminates at various temperatures in the presence of moisture. Such experimental evaluation requires a knowledge of the moisture diffusion process in order to perform moisture conditioning and to determine moisture content and distribution during elevated temperature tests.

5.2.1 Moisture Diffusion in a Thin Composite Laminate

In most engineering applications, moisture diffusion is through a large surface area with very few edges. As a result, diffusion through the thickness is of primary interest. For such a case, thin composite specimens are utilized in order to approximate a one-dimensional diffusion process. If the use of a thin composite is not practical, the edges can be sealed with foil or an appropriate coating to retard diffusion.

It has been previously shown[3] that moisture diffusion in laminated composites can be predicted by Fick's Second Law. For one-dimensional diffusion through the thickness of an infinite plate of constant thickness, h, the diffusion process is described by the relationship[7]

$$\frac{\partial c}{\partial t} = \overline{d}\,\frac{\partial^2 c}{\partial z^2} \tag{5.1}$$

where c is the moisture concentration, $\overline{d}$ is the effective diffusivity through the thickness, t denotes time, and z is the thickness coordinate. The term effective is used in conjunction with the diffusivity because we are dealing with a laminated system in which the diffusivity varies from ply-to-ply for the general case.

For a laminate initially in a dry state, a solution to eq (5.1) in terms of percent weight gain, M, can be approximated by the following expressions[8,9]

$$G = 4\sqrt{\frac{t^*}{\pi}} \qquad t^* < 0.07 \tag{5.2}$$

$$G = 1 - \frac{8}{\pi^2}\exp(-\pi^2 t^*), \qquad 0.07 \le t^* < 1 \tag{5.3}$$

$$G = 1, \qquad 1 \le t^* \tag{5.4}$$

where

$$G = \frac{M}{M_e} \tag{5.5}$$

$$t^* = \frac{\overline{d}t}{h^2} \tag{5.6}$$

and M_e is the percent weight gain associated with equilibrium. Thus, a plot of M versus $\sqrt{t}$ yields a straight line over the initial portion of the curve. The slope of the linear region is directly related to $\overline{d}$, i.e.,

$$\bar{d} = \frac{\pi h^2}{16} \left(\frac{M}{M_e \sqrt{t}}\right)^2 \tag{5.7}$$

Equation (5.7) provides a relationship for experimentally determining $\bar{d}$.

5.2.2 Moisture Measurements

Diffusion tests usually consist of measuring weight gain as a function of time for a constant temperature and humidity exposure. The concentration of moisture on the surface of an exposed specimen is a function of relative humidity, and as a result M_e is also a function of humidity.

Diffusion coefficients are a function of temperature and appear to follow an Arrhenius type relationship of the form[3,10]

$$d = d_o \exp(-E_d/RT) \tag{5.8}$$

where d_o is a constant, E_d is the activation energy for diffusion, R is the universal gas constant, and T is temperature as measured on a Kelvin scale. Thus, the diffusion process must be characterized for different temperatures as well as relative humidities.

All specimens for absorption should be pre-conditioned in a vacuum oven until a near-equilibrium weight is obtained. This procedure will assure initially dry specimens. The temperature of the vacuum oven should not exceed 200° F (93° C). The drying temperature should never exceed T_g. Thus, for resins with low values of T_g, the vacuum oven temperature may have to be less than 200° F (93° C).

After pre-conditioning, dry specimens are placed in an environmental chamber under constant temperature and constant humidity. If a chamber in which relative humidity can be controlled is not available, specimens can be placed in a container of water inside an oven, and the underwater condition will approximate 100 percent relative humidity. If other humidity conditions are desired, a humid dessicator can be employed in conjunction with the oven.[11] Specimens are removed from the chambers at various time intervals for weight measurement. The effect of this removal on weight gain determination has been shown to be negligible.[10] Specimens should be allowed to cool for a short period of time before being weighed. For specimens placed underwater, any surface moisture should carefully be wiped off before weighing. The ideal exposure, obviously, would consist of a chamber in which weight gain could be monitored continuously without specimen removal. The weight gain process is continued until an apparent equilibrium is reached.

The data reduction procedure consists of plotting M/M_e as a function of $\sqrt{t}$ for each exposure temperature. Since the data are normalized by M_e, different humidity conditions can be plotted on the same curve. For thin specimens the data should approximate a straight line in the region $0 \geq M/M_e \leq 0.6$. The slope of the straight line portion of the M/M_e versus $\sqrt{t}$ plot is used in conjunction with eq (5.7) to obtain $\bar{d}$.

If desorption experiments are to be utilized for measuring diffusivity, the procedures as described here require that moisture absorption must occur until equilibrium is reached prior to running the desorption test. This assures an initial moisture profile that is uniform. If the initial moisture distribution is not uniform, the data reduction procedures described are not valid. For desorption $M_e = 0$ and eq (5.5) becomes

$$G = \frac{M_0 - M}{M_0} \tag{5.9}$$

where M_0 is the initial equilibrium moisture content. Thus, the previously described procedures can be used for desorption except the expression in eq (5.9), rather than M/M_e, is plotted as a function of $\sqrt{t}$. A more complete discussion of combinations of absorption and desorption experiments can be found in Ref. 11.

Note that all of the diffusion measurements discussed consider only total weight gain. To date no acceptable method has been developed for measuring the moisture distribution for composite materials.

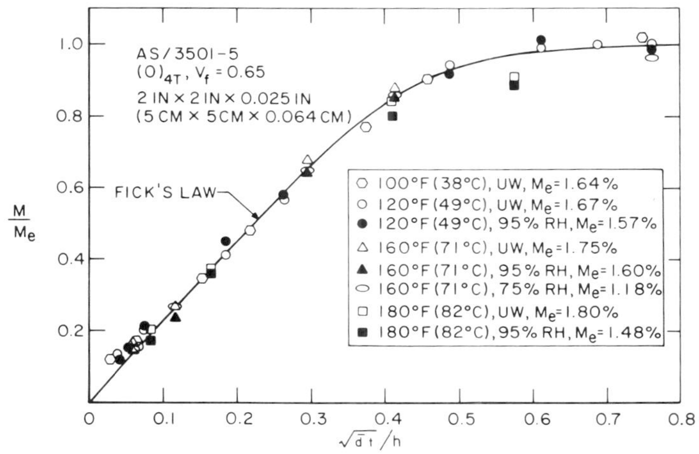

Fig. 5-4—Master plot for graphite/epoxy unidirectional composite (Whitney and Browning[10])

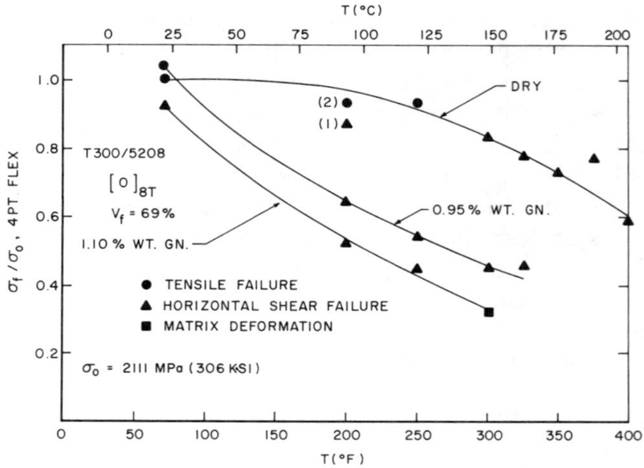

Fig. 5-5—Flexure strength, four-point loading of T300/5208 graphite/epoxy unidirectional composites (Whitney and Husman[12])

5.3 MECHANICAL PROPERTIES

In most engineering usage of fiber reinforced composites, laminate stacking geometry is chosen such that stiffness and strength are controlled by fiber modulus and strength, respectively. Thus, some matrix softening can be accommodated in such applications without serious consequences. If considerable matrix softening occurs, however, the ability of the resin to support the fiber is severely reduced, along with the ability to transfer load through the matrix to the fibers. The result is a change in failure mode from filament dominated to matrix dominated. The classical example is that of unidirectional compression, where a significant loss in matrix stiffness leads to local instabilities and a reduction in compression strength. Thus, any loss in resin T_g due to moisture absorption can lead to a reduction in the useful temperature range of the composite laminate.

Another classic example of a failure mode change is the unidirectional flexure test. These tests are commonly used for quality control, and 0-deg flex strength is considered to be a filament dominated property. For state-of-the-art high performance epoxy resins, 0-deg dry flex strength is relatively insensitive to temperature below 300° F (149° C). With increasing moisture content, however, substantial strength degradation can occur at temperatures considerably below 300° F (149° C). Such strength degradation is accompanied by a change in failure mode from filament dominated to matrix dominated.[12] This is illustrated in Fig. 5-5 for four-point loading of a graphite-epoxy unidirectional composite.[12]

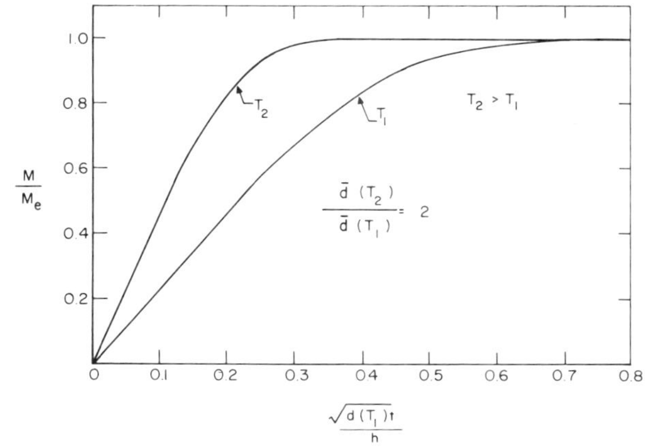

Fig. 5-6—One dimensional moisture absorption for temperatures T_1 and T_2

Thus, it is easily seen from these examples that the characterization of mechanical properties in the presence of moisture and temperature is of practical concern.

5.3.1 Accelerated Conditioning

In order to minimize pre-conditioning time, accelerated moisture conditioning techniques are often employed in conjunction with mechanical test specimens. This can be accomplished by increasing exposure temperature and humidity.

As previously noted, the value of G in eqs (5.2–5.4) is determined by the value of t^*. A cursory examination of eq (5.6) reveals that t can be decreased for given values of t^* and h by increasing $\bar{d}$. This can be accomplished by increasing exposure temperature in accordance with eq (5.8). This is illustrated in Fig. 5-6 where eqs (5.2–5.4) are plotted for one-dimensional diffusion in conjunction with two different diffusivities, $\bar{d}(T_1)$ and $\bar{d}(T_2)$, associated with temperatures T_1 and T_2.

If the pre-conditioning moisture content is less than the value of M_e associated with 100 percent humidity exposure, then exposure at high humidity levels will accelerate the conditioning time. For example, if M_e = 2.0 percent for 100 percent humidity and 1.0 percent for 50 percent relative humidity, then the equilibrium moisture content associated with the lower relative humidity can be obtained by exposure at a humidity level above 50 percent for a shorter period of time. This is illustrated in Fig. 5-7 where eqs (5.2–5.4) are plotted for one-dimensional diffusion. Actual weight gain, M, is shown for both examples of 100 percent and 50 percent relative humidity exposure. It is obvious that

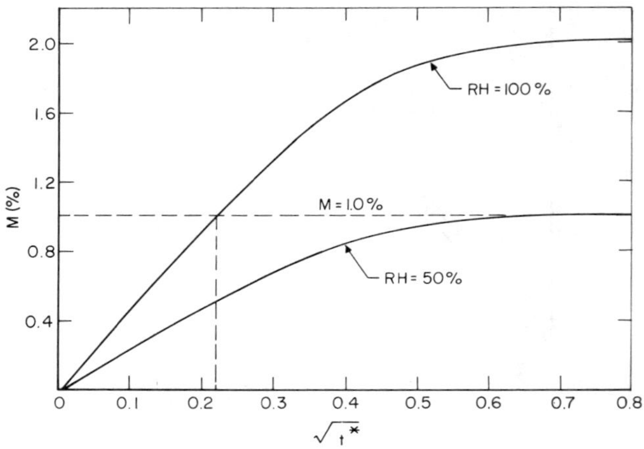

Fig. 5-7—One dimensional weight gain for two different relative humidity exposures

considerable time can be saved in obtaining the 1.0 percent weight gain by exposure at 100 percent relative humidity. It should be noted, however, that this procedure yields a very non-uniform moisture distribution across the laminate thickness for the 1.0 percent weight gain associated with 100 percent relative humidity. The moisture profiles associated with 1.0 percent weight gain for humidity exposures are shown in Fig. 5-8 with the local percent weight gain per unit thickness, *m*, determined from the relationship[7,8]

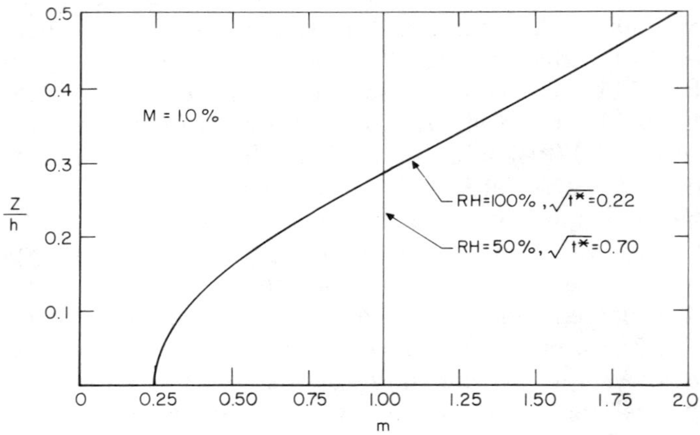

Fig. 5-8—Moisture profiles at 1.0 percent weight gain for two relative humidity exposures

$$m = \frac{M_e}{h} \left[1 - erf\left(\frac{1 - 2z/h}{4\sqrt{t^*}}\right)\right] \qquad t^* < 0.01 \tag{5.10}$$

$$m = \frac{M_e}{h} \sum_{k=1}^{3} (-1)^{k+1} \left[erfc\left(\frac{2k - 1 - 2z/h}{4\sqrt{t^*}}\right) + erfc\left(\frac{2k - 1 + 2z/h}{4\sqrt{t^*}}\right)\right],$$

$$0.01 \le t^* < 1 \tag{5.11}$$

$$m = \frac{M_e}{h}, \qquad 1 \le t^* \tag{5.12}$$

where z is measured from the center of the composite.

Note that while the total weight gain M is at the desired 1.0 percent level for 100 percent exposure, the outer surface, $z/h = 0.5$, is at 2.0 percent, and the midplane, $z/h = 0$, is almost dry. Such a gradient may yield very misleading results in terms of mechanical properties.

Care should also be taken in high-temperature moisture exposure. in particular, temperatures which are close to the wet glass transition temperature of the resin may induce permanent damage to the test specimen in the form of cracks. In addition, if large amounts of water associated with high-humidity exposure are rapidly forced into the composite by high temperatures, rapid swelling will occur with an increased possibility of inducing cracks. Further discussion on these points can be found in Refs. 10 and 11.

5.3.2 End Tabs and Strain Gages

Most of the standard adhesives used for strain gages and specimen end tabs require heat and pressure for varying lengths of time. Such procedures induce complete drying of pre-conditioned specimens. Moisture sensitivity of most adhesives precludes bonding prior to moisture conditioning. Silicone rubber and foil have been placed over strain gages in order to isolate them from moisture contact during environmental conditioning. Such a procedure is only partially effective, however, as moisture diffusion from inside the composite to the outer surfaces will eventually penetrate the bondline.

For room-temperature specimens these problems can be overcome by utilizing a room-temperature curing, anaerobic adhesive (for example, Eastman 910 or M-Bond 200). Such adhesives can be used for end tabbing and strain gaging after moisture conditioning. For elevated temperature specimens, a high-temperature anaerobic adhesive (for example, Loctite 306/NF Primer) can be used. This type of adhesive will set up at room temperature and complete its cure during heating of the specimen in preparation for testing. This procedure as applied to strain gages is fully described in Ref. 13.

5.3.3 Moisture Control

Very often it is necessary to perform mechanical tests at elevated temperatures in a dry oven which induces complete drying of the specimen. Although environmental chambers are available which control humidity below the boiling point of water, higher temperatures with humidity control cannot be obtained without a pressure chamber. Such equipment is not commonly found in the laboratory.

As previously discussed in section 5.2.2, the diffusivity is a function of temperature. Thus, the amount of weight loss associated with elevated temperature tests in a given time period will be a function of temperature. In order to compare the degradation in mechanical properties at various elevated temperatures for a given moisture content, the hold time prior to testing must be adjusted for different temperatures.

Such a procedure was implemented in Ref. 12 in conjunction with unidirectional flexure specimens. It can easily be seen from eqs (5.2-5.4) and eqs (5.10-5.12) that both the total weight gain and detailed moisture profile are determined by t^* for thin composites. Thus, control of t^* provides a means of controlling the detailed moisture distribution during a drying process. In particular, a plot of M versus the $\sqrt{t}$ should yield a straight line for a process in which the diffusion coefficient is constant. From such a plot, a value of t can be chosen for different temperatures such that t^* is a constant. This will assure the same total weight loss and moisture profile for each temperature associated with a given value of M_0.

In order to control the moisture content during the flexure experiments of Ref. 12, weight loss at various temperatures as a function of time was determined for specimens fabricated from two different material systems. Results are shown in Fig. 5-9 for AS/3501-5 graphite/epoxy specimens. The test procedures consisted of pre-heating the test chamber, including the flexure fixture, to the desired temperature, then placing the specimen in the loading fixture, and holding a predetermined length of time before loading the specimen. A minimum time of about 90 seconds was required to bring the specimen to temperature equilibrium. The weight loss at all elevated temperatures was determined by the 163° C (325° F), 90-second weight loss. This was the maximum test temperature. For the AS/3501-5 specimens in Fig. 5-9, this required a seven-minute temperature soak at 121° C (250° F) and 22-minute soak at 93° C (200° F).

The data in Fig. 5-9 are plotted in Fig. 5-10 as a function of $\sqrt{t}$. For the 93° C (200° F), an approximate straight line is obtained. For the

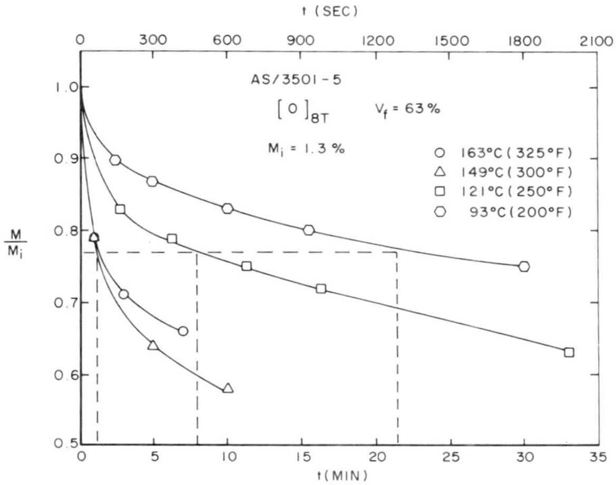

Fig. 5-9—Weight loss function of time, AS/3501-5 unidirectional flexure specimens (Whitney and Husman[12])

other elevated temperatures, transient heating effects in conjunction with the shorter times yield results which depart drastically from linearity. Thus, the moisture profile was not the same for all elevated temperatures, although the total moisture content at the beginning of the test was the same.

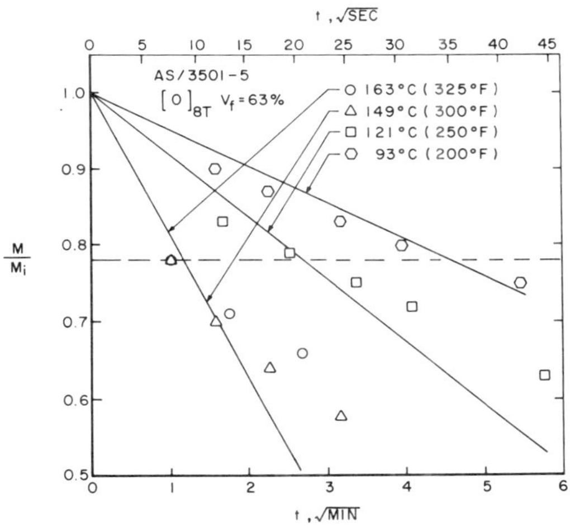

Fig. 5-10—Weight loss as a function of square root time, AS/3501-5 unidirectional flexure specimens (Whitney and Husman[12])

Because of these difficulties in controlling weight loss during elevated temperature tests, it is strongly recommended that mechanical testing be carried out in a humidity controlled environment whenever possible.

REFERENCES

1. Bueche, F., *Physical Properties of Polymers,* Interscience Publishers, New York (1962).
2. McKague, E.L., Jr., Halkias, J.E. and Reynolds, J.D., "Moisture Diffusion in Composites: The Effect of Supersonic Service on Diffusion," *Journal of Composite Materials,* Vol. 9, 2-9 (1975).
3. Shen, Chi-Wung and Springer, G.S., "Moisture Absorption and Desorption of Composite Materials," *Journal of Composite Materials,* Vo.. 10, 2-20 (1976).
4. McKague, E.L., Jr., Reynolds, J.D. and Halkias, J.E., "Moisture Diffusion in Fiber Reinforced Plastics," *ASME Journal of Engineering Materials and Technology,* Vol. 98, Series H, 92-95 (1976).
5. Browning, C.E., Husman, G.E. and Whitney, J.M., "Moisture Effects in Epoxy Matrix Composites," *Composite Materials: Testing and Design (Fourth Conference),* ASTM STP 617, American Society for Testing and Materials, 481-496 (1977).
6. Bueche, F. and Kelley, F.N., "Viscosity and Glass Temperature Relations for Polymer-Diluent Systems," *Journal of Polymer Science,* Vol. 45, 267-273 (1960).
7. Crank, J., *Mathematics of Diffusion,* Second Edition, Oxford University Press (1975).
8. Whitney, J.M., "Moisture Diffusion in Fiber Reinforced Composites," *Proceedings of the Second International Conference on Composite Materials,* The Metallurgical Society of AIME, 1584-1601 (1978).
9. Whitney, J.M., "Three-Dimensional Moisture Diffusion in Laminated Composites," *AIAA Journal,* Vol. 15, No. 9, 1356-1358 (Sept. 1977).
10. Whitney, J.M. and Browning, C.E., "Some Anomalies Associated with Moisture Diffusion in Epoxy Matrix Composite Materials," *Environmental Effects on Advanced Composites,* ASTM STP 658, American Society for Testing and Materials, Philadelphia (1978).
11. Shirrell, C.D., "Diffusion of Water Vapor in Graphite/Epoxy Composites," *Environmental Effects on Advanced Composites,* ASTM STP 658, American Society for Testing and Materials, Philadelphia (1978).
12. Whitney, J.M. and Husman, G.E., "Use of the Flexure Test for Determining Environmental Behavior of Fibrous Composites," *Experimental Mechanics,* Vol. 8, No. 5, 185-190 (May 1978).
13. Fowler, C.C., Jr., "Bonding of Elevated Temperature Strain Gages to Humid Aged Graphite Tensile Specimens Through the Use of Anaerobic Adhesives," Air Force Technical Report *AFML-TR-75-204,* Air Force Materials Laboratory (Dec. 1975).

INDEX